Antenna Arrays - Applications to Modern Wireless and Space-Born Systems

Edited by Hussain M. Al-Rizzo, Nijas Kunju, Sulaiman Tariq and Aldebaro Klautau

Published in London, United Kingdom

Antenna Arrays - Applications to Modern Wireless and Space-Born Systems
http://dx.doi.org/10.5772/intechopen.100726
Edited by Hussain M. Al-Rizzo, Nijas Kunju, Sulaiman Tariq and Aldebaro Klautau

Contributors
Habib Mohammed, Tessema T. Terefe, Sultan Feisso, M. Rani Laxmi Prasanna Rani, Moturi Satyanarayana, Narsupalli Shanmukha Rao, Smrity Dwivedi, Hua Li, Zhenning Li, Kaiyu Liu, Mingshan Ren, Yunkai Deng, Onu Kingsley Eyiogwu, Hussain Al-Rizzo, Ken G. Clark, Jim M. Tranquilla

First published in London, United Kingdom, 2022 by IntechOpen
IntechOpen is the global imprint of INTECHOPEN LIMITED, registered in England and Wales, registration number: 11086078, 5 Princes Gate Court, London, SW7 2QJ, United Kingdom

British Library Cataloguing-in-Publication Data
A catalogue record for this book is available from the British Library

Additional hard and PDF copies can be obtained from orders@intechopen.com

Antenna Arrays - Applications to Modern Wireless and Space-Born Systems
Edited by Hussain M. Al-Rizzo, Nijas Kunju, Sulaiman Tariq and Aldebaro Klautau
p. cm.
Print ISBN 978-1-80356-404-3
Online ISBN 978-1-80356-405-0
eBook (PDF) ISBN 978-1-80356-406-7

Meet the editors

Hussain Al-Rizzo received his Ph.D. in Electrical and Computer Engineering Department from the University of New Brunswick, Canada. In 2000, he joined the Systems Engineering Department at the University of Arkansas at Little Rock where he is currently a Professor of Telecommunication Systems Engineering. He has published more than 300 papers in peer-reviewed journals and conference proceedings, two books, eight book chapters, and four patents. His research areas include V2V, V2X, and V2I wireless systems; smart antennas; massive MIMO; flexible RF components and antennas; implantable medical devices; electromagnetic wave scattering by complex objects; design, modeling, and testing of high-power microwave applicators; precipitation effects on GPS; terrestrial and satellite frequency re-use communication systems; field operation of NAVSTAR GPS receivers; data processing and accuracy assessment; and effects of the ionosphere, troposphere, and multipath on code and carrier-beat phase GPS observations.

Aldebaro Klautau received a bachelor's (Universidade Federal do Pará, UFPA, 1990), M. Sc. (Universidade Federal de Santa Catarina, UFSC, 1993), and Ph. D. degrees (University of California at San Diego, UCSD, 2003) in Electrical Engineering. Since 1996, he has been with UFPA and is now a full professor, at the ITU Focal Point, and directs LASSE. He was a visiting scholar at Stockholm University, UCSD, and The University of Texas at Austin. He is a researcher of the Brazilian National Council of Scientific and Technological Development (CNPq), a senior member of the IEEE, and a senior member of the Brazilian Telecommunications Society (SBrT). He has supervised more than 60 graduate students, published more than 200 papers in peer-reviewed conferences and journals, and has several international patents.

Sulaiman Tariq received his Ph.D. in Telecommunication Systems Engineering from the University of Arkansas at Little Rock in 2022 and his B.Sc. in Electrical and Electronics Engineering from Khalifa University in Abu Dhabi, UAE in 2017. He is currently working as a Wireless Systems Engineer with Apple Inc. in Cupertino Silicon Valley, California. He has published two IEEE and IET journal papers and two book chapters. Sulaiman's research areas include miniaturized antenna arrays design, RF circuits, MIMO/Massive MIMO, 5G, Wi-Fi-6, V2X communications, ray-tracing and stochastic channel modeling.

Nijas Kunju received an MSc (2009) and Ph.D. (2015) in Electronics Science from Cochin University of Science and Technology, Cochin, Kerala under the supervision of Prof. P Mohanan. He is currently working as Technical Manager at ANSYS Inc. He is specialized in the domain of RF, Antennas, Radar, Signal Integrity, Power Integrity, and Multiphysics simulation. Before he joined ANSYS, he was working as Research Consultant at GE Global Research in Bangalore in the area of microwave material sensing for detecting different materials in Crude Oils. He is holding 20+ International publications and 20+ national and international conference publications.

Contents

Preface

Modern communications systems require the use of multiple antennas to meet key performance indicators. In particular, 5G and 6G mobile communications rely on the existence of antenna arrays to achieve a wide variety of desired use cases. Several antenna designs are evolving to improve systems that incorporate Multiple-Input Multiple-Output (MIMO), massive MIMO, cell-free MIMO, and other techniques. The expected convergence of terrestrial and non-terrestrial networks also requires research on antennas and antenna arrays. In this context, this book presents recent developments in modern design and analysis of antenna array techniques. It consists of six chapters and is useful for interested readers and those working with antenna arrays, including researchers, engineers, and students.

Chapter 1 presents the design and analysis of three-dimensional GPS microstrip antennas operating at 1.57542 GHz. The simulations are compared with experimental results and show the impact of several 3D geometries on figures of merit such as the phase response. The proposed antenna presents a nearly uniform phase response when the 3D structure is bent by angles ranging from 15 degrees to 30 degrees without compromising performance. A Finite-Difference Time-Domain scheme is discussed to clarify the interplay between the proposed antennas and several ground plane configurations.

Chapter 2 discusses present and future research and development directions regarding spaceborne Synthetic Aperture Radar (SAR) antenna arrays. It includes theoretical analysis, simulation results, and experimental verification. The authors explain the drawbacks of traditional antenna arrays, which include large volumes, weights, and manufacturing costs, among others. Then some breakthroughs in antenna architectures are presented, such as the metamaterial and periodic reflector array antennas. The chapter also discusses aperiodic, parabolic, and hypersurface antennas, indicating characteristics and trends of spaceborne SAR antenna arrays.

Chapter 3 describes a method to design antenna arrays using machine learning. Within the context of array thinning, the goal is to systematically remove elements without change in the performance. The presented results include a performance evaluation of arrays with and without thinning and the radiation characteristics. The authors also discuss the characteristics of deep learning and other learning algorithms in the design of antenna arrays with thinning.

Chapter 4 investigates the reduction of the sidelobe level associated with antenna arrays using enhanced firefly and genetic algorithms. The presented results show that the enhanced firefly algorithm outperforms the genetic algorithm. The antenna array side lobe level is optimized without degrading performance with respect to the beam width. The author also discusses the advantages of reducing the number of active elements in an antenna array by turning them off.

Chapter 5 presents an analysis of coverage considering pathloss models adequate for studying digital terrestrial TV and TV white space. It analyzes different propagation models in very high and ultra-high frequency bands. Then, the authors use the best model to find the coverage of the incumbent transmitter and the free channels for secondary use.

Chapter 6 focuses on the development of multipurpose antennas having different applications in communications and detection, among other applications. It discusses techniques for antenna array reconfiguration, with special emphasis on graphene antennas. The chapter includes a few antenna array designs, which help in understanding reconfiguration for distinct applications.

Hussain M. Al-Rizzo
Department of Systems Engineering,
DSTEM College,
University of Arkansas Little Rock,
USA

Nijas Kunju
Ansys Inc.,
Bangalore, India

Aldebaro Klautau
Computer Engineering Department,
Federal University of Para,
Belem, Brazil

Sulaiman Tariq
Wireless Systems Engineer (Apple Inc.),
Cupertino, USA

Chapter 1

Three-Dimensional Microstrip Antennas for Uniform Phase Response or Wide-Angular Coverage GPS Applications

Ken G. Clark, Jim M. Tranquilla and Hussain M. Al-Rizzo

Abstract

In this chapter, three-dimensional microstrip antennas operating in the dominant TM_{10} mode of the L1 band at 1.57542 GHz are presented, which either provide uniform phase response and stable phase center, or wide angular beam coverage. A Finite-Difference Time-domain scheme has been developed and used to elucidate the interplay between the proposed antennas and several ground plane configurations. Experimental results are presented and compared to simulations to illustrate the impact of several 3-D geometries on the phase response, phase center, half power beam width, gain, and polarization purity. We report for the first-time novel three-dimensional microstrip antenna topologies for precise surveying applications by fostering a nearly uniform phase response when the 3D structure is bent by angles ranging from 15° to 30° **without significantly compromising the gain and polarization performance** of the equivalent flat patch antenna. The proposed antennas exhibit half-power-beamwidths up to 40 and 80% greater than the equivalent flat structure that can be realized if positioned downward and upward respectively, for bend angles above 45° accompanied with a variety of spatial coverage patterns to suit applications involving highly dynamic platforms.

Keywords: three-dimensional (3D) microstrip antennas, global positioning system, low-elevation pattern coverage, GPS tracking of sounding rockets and unmanned aerial vehicles, GPS aerospace applications

1. Introduction

The design requirements of a global positioning system (GPS) antenna rely on the application under consideration. A GPS user antenna is required to exhibit right-hand circular polarization (RHCP) and RHCP radiation pattern coverage from zenith down to five elevations for all azimuth angles to maintain tracking of the full visible satellite constellation during dynamic maneuvers [1–7]. An equally important, yet more challenging requirement is that the antenna must provide a virtually uniform phase response and stable phase center over the coverage region to satisfy precise positioning accuracy requirements [8–10].

Due to their light weight, size, low cost, and integration with printed-circuit boards, microstrip antennas offer an attractive solution to meet some of these design requirements. A considerable amount of research has been reported on the design of microstrip antennas for commercial GPS applications [11–16]. It is well known that conventional planar microstrip antennas suffer from reduced gain at low-elevation angles [7] which subsequently lead to loss of contact with GPS-received signals in particular when installed on highly-dynamic vehicles.

The demand for antennas that provide horizon-to-horizon coverage has risen considerably to keep pace with the recent stringent needs of modern GPS marine navigation and aerospace applications [16–21]. In this paper, we present three-dimensional microstrip antennas (3DMAs) for maritime GPS applications whereby pitch and roll amplitudes as high as 10–15° are encountered during adverse weather conditions [2]. Hence, the GPS antenna must provide coverage extending to negative elevation angles to compensate for vessel motion due to pitching and rolling [17–19]. Other potential applications for the antennas presented in this paper include tracking of unmanned aerial vehicles, space-borne vehicles, and sounding rockets where a key design objective is to provide a quasi-static isotropic radiation pattern to acquire the locally visible GPS satellites under the tumbling motion, during ascent trajectory and re-entry into the dense atmosphere [20, 21].

A few 3DMAs designs have been proposed to improve the radiation pattern of planar microstrip antennae in order to suit GPS applications involving highly-dynamic vehicles. A GPS manufacturer patented a downward 3DMAs design [5]. However, it should be noted that neither the dimensions were disclosed nor the claimed performance metrics were quantified in [5]. A corner truncated square patch, mounted on a pyramidal ground plane and partially enclosed within a flatly folded conducting wall was reported in [6] with a 3-dB Axial Ratio (AR) beamwidth of 130°. However, neither the phase response nor the phase center characteristics reported in [6]. In [7], we proposed a downward drooped square annular element resonating in the TM_{30} mode at 1.57542 GHz which revealed a complete upper hemispherical coverage with the pattern ripple being reduced to less than 2 dB. However, the antenna suffers from increased size (18.6 cm square annular element printed on a 4-mm thick substrate with a relative permittivity, $\varepsilon_r = 2.2$). Moreover, the cross-polarized component became dominant by 1.5 dB near the horizon and the phase pattern displayed large variations in the elevation cut.

In contrast to the antennas reported in [5–7], this paper presents novel 3DMAs operating in the fundamental TM_{10} resonance mode, in which the ground plane and patch element are bent either downward or upward such that the corners or edges of the resonant cavity region fall away from the plane occupied by the patch. It should be emphasized that a fundamental understanding of the benefits and limitations of 3DMAs is still lacking. The key design parameters which influence the performance of the phase response of the far-field radiation pattern of 3DMAs such as bend angle, length of the flat portion, size of the ground plane, and substrate have not yet been reported in the literature.

In this paper, we present several downward and upward 3DMAs which are designed and characterized using a rigorous full-wave electromagnetic model and experimentally validated via measurements conducted inside an anechoic chamber. Our aim in investigating the 3DMAs is twofold: (i) to provide uniform phase response along with a stable phase center over the entire coverage region, and (ii) to symmetrically cover much of the upper hemisphere and to extend coverage to negative elevation angles so that the receiver can maintain lock with sufficient signal-to-noise ratio for all angles of the desired view. The first design goal is critical for minimizing carrier and group delay variations in phase-tracking receivers. The second prevents occurrences of cycle slips and loss of lock to satellites, thereby reducing the Root-Mean-Square (RMS) error in position, velocity, acceleration, attitude, roll, pitch, and yaw of a moving platform [2, 18, 19].

The rest of the chapter is organized as follows. In Section 2, the geometries of the antennas under consideration are introduced. Section 3 summarizes the CPFDTD algorithm developed to perform the design and parametric studies along with experimental results to demonstrate its ability to correctly predict the resonant frequency and far-field radiation patterns. Section 4 introduces the antennas designed, constructed and tested for the control of the far-field radiation patterns with special emphasis on phase performance. Section 5 concludes the Chapter.

2. The design of the 3DMAs geometries

The reference Flat Microstrip Antenna (FMA), shown in **Figure 1(a)**, consists of a square copper patch, L_p in side length. The dielectric substrate has a thickness of h and the relative permittivity is denoted by ε_r. The patch is rotated by 45 with respect to the substrate layer.

Two types of 3DMAs are considered in this Chapter. The first is bent downward D3DMA shown in **Figure 1(b)**, the cross-sectional view of which is shown in **Figure 1(c)**. The second is the bent upward U3DMAs depicted in **Figure 2(a)**. The D3DMA is shown in **Figure 2(b)**. The side lengths of the ground plane and substrate are denoted by $L_g \times L_g$ and $L_s \times L_s$, respectively, where $L_g = L_s$. The D3DMA has a square flat top with a side length of L_{ft}, and the drooped angle with respect to the flat top side is denoted by ϕ. The length, L_p of the square patch is chosen to

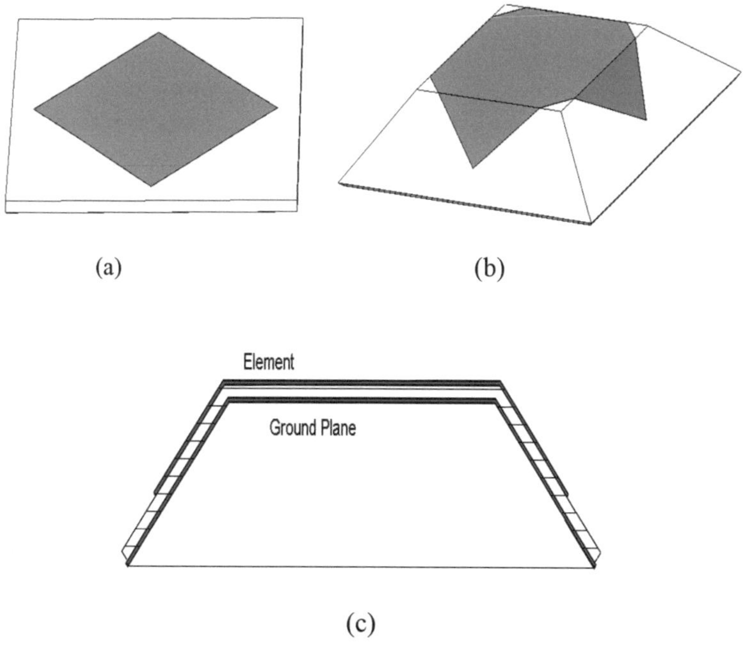

Figure 1.
(a) Reference FMA, (b) the D3DMA, (c) The resonant cavity.

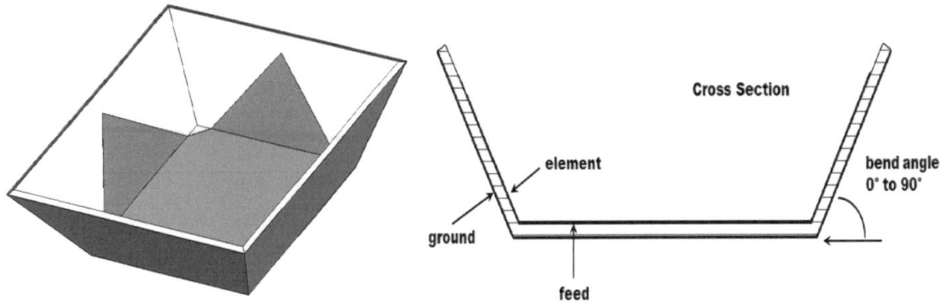

Figure 2.
(a) The U3DMAs, (b) The resonant cavity.

achieve the desired resonant frequency and bandwidth of the GPS L_1 carrier and varies according to ϕ, h, and ε_r. A 50 Ω coaxial probe is used to feed all antennas presented in this Chapter, the inner conductor terminates on the patch.

3. Experimental validation of the CPFDTD modeling approach

The full-wave CPFDTD technique has been employed due to its flexibility for modeling complex structures. A versatile 3-D CPFDTD algorithm, combined with a thin-wire approximation, is implemented is developed with the formulation whereby different cell sizes can be chosen along the three orthogonal axes. Geometrical symmetry is implemented using the Neumann boundary condition. The input impedance of the antenna is obtained by exciting the coaxial cable port using a Gaussian pulse where the width and delay are determined from the resonant frequency and desired bandwidth. The appropriate electric and magnetic field components are integrated at the feed port to yield the voltage across and current flowing in the inner conductor which subsequently allows the computation of the input impedance in the frequency domain using Fourier transformation. To evaluate the magnitude and phase of the far-fields a sinusoidal signal where the frequency is chosen at the resonant mode of the structure and the simulations are ramped up to steady-state value over a number of cycles to ensure convergence to the desired accuracy.

The CPFDTD model is verified against the measured resonant frequency and ar-field radiation patterns. A simple prototype is simulated and constructed using only two edges bent down by an angle of 45° as shown in **Figure 3**.

These electromagnetic boundary conditions are enforced on the PEC conducting patch and ground plane which are not oriented along the Cartesian planes as follows. The electric and magnetic field components which are adjacent to the interface between two boundaries are updated using an integral approach applied to the fields that surround the contour of the appropriate cell [22]. This approach is better suited to conform to the 3-D surfaces under consideration to reduce numerical dispersion caused by traditional stair-step FDTD approximation.

As shown in **Figure 4**, the drooped surfaces are discretized using three different meshing schemes to suit a given bend angle. The vertical to horizontal mesh steps Δx and Δz are varied to conform to the drooping surface.

The measured resonant frequency is 1.55 GHz as compared to 1.575 GHz obtained from the CPFDT model. The measured phase response in the E-plane is displayed in **Figure 5**. The far-field patterns reveal asymmetry which is caused by the offset in the

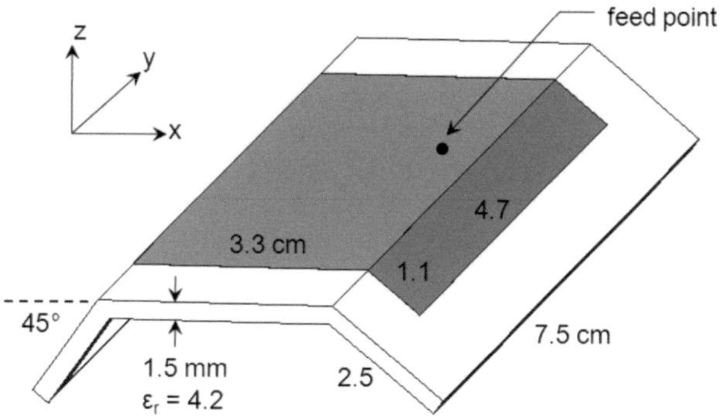

Figure 3.
Prototype with two drooped edges; all dimensions are in cm.

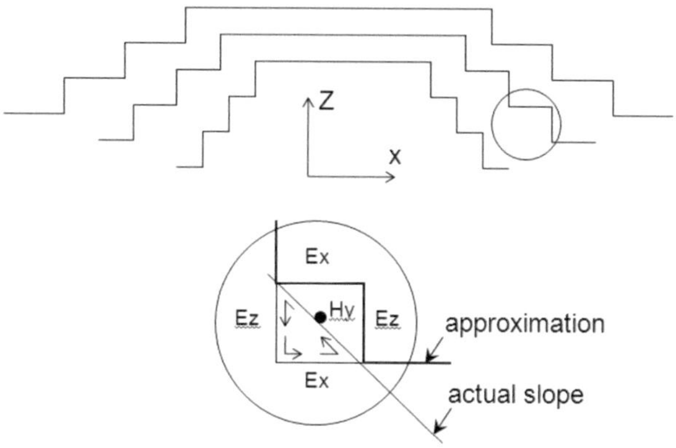

Figure 4.
Approximation of shallow, intermediate, and steep surfaces.

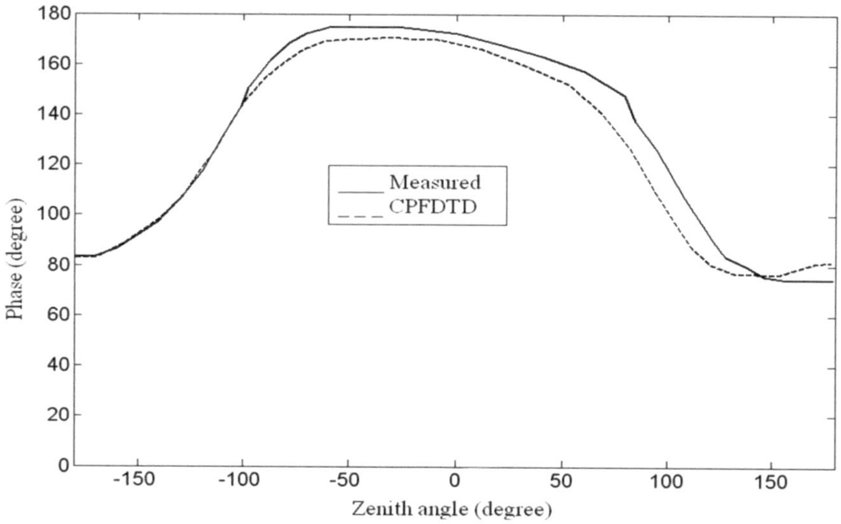

Figure 5.
Measured and calculated phase pattern in the E-plane (x-z plane) for the antenna shown in ***Figure 4****.*

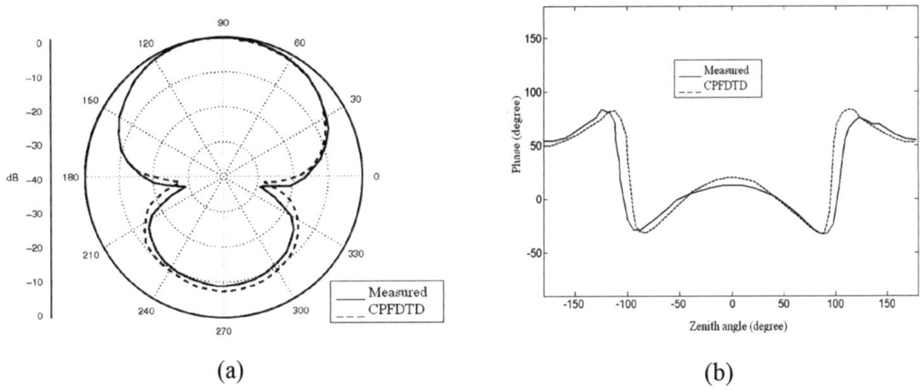

Figure 6.
(a) Amplitude and (b) phase of the measured versus simulated far-field patterns in the H plane.

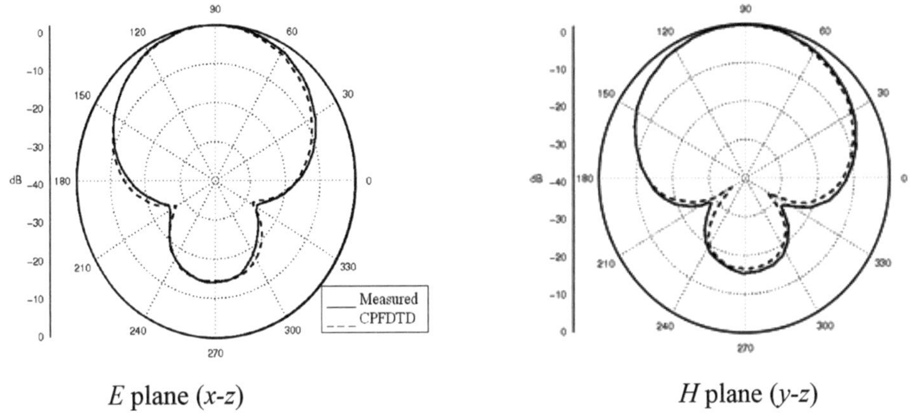

Figure 7.
Measured and simulated elevation patterns for the 30° drooped antenna, ε_r = 2.2.

antenna mount inside the anechoic chamber which is necessary in order to install the bends and cable on the antenna mount. For comparison, we referenced the calculated far-field patterns to the same offset origin. Results shown in **Figure 5** along with the amplitude and phase of the normalized H-plane patterns displayed in **Figure 6** show good agreement between measured and simulated results which validates our simulation approach.

Next, the bends are duplicated on the remaining two sides to arrive at the D3DMA geometry. We constructed a downward DMA to experimentally validate the final CPFDTD model. The prototype consists of a 62 × 62 mm patch, L_f = 40 mm, ϕ = 60° printed on a substrate with L_s = 100 mm, h = 1.5 mm, and ε_r = 4.2. As shown in **Figure 7**, excellent agreement is seen between the simulated and measured far-field radiation patterns in the E- and H-planes.

4. Design procedure and parametric study

In this section the impact of the bend angle, ϕ, length of the flat top, L_f, size of the ground plane, L_g, height, h, and the permittivity of the substrate, ε_r are analyzed to

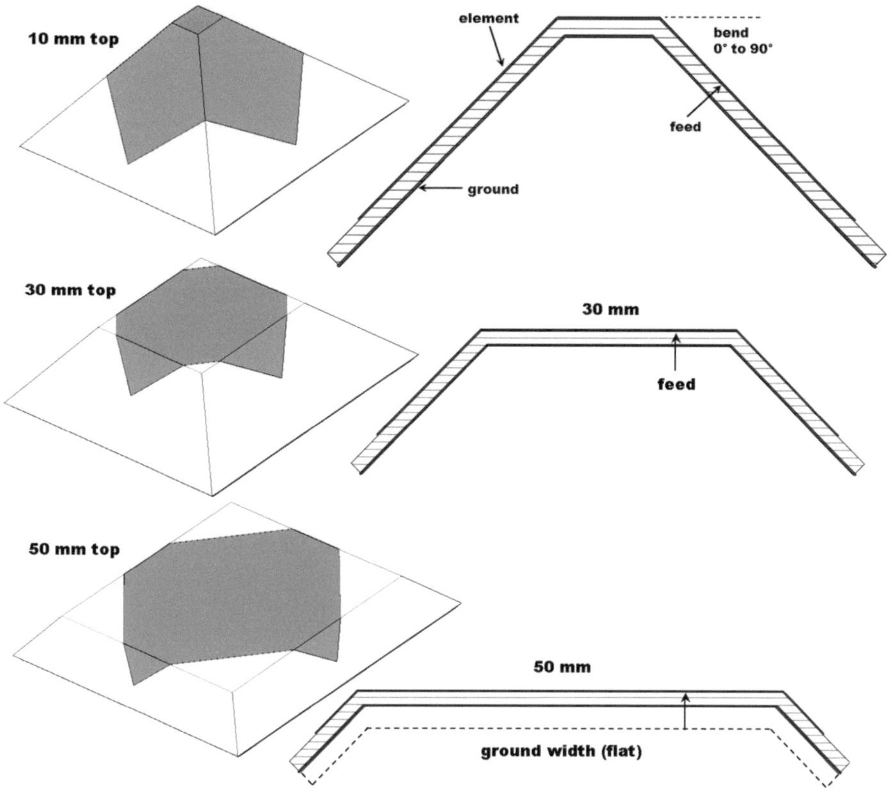

Figure 8.
Structural variations: L_f *= 10, 30, and 50 mm;* L_g *= 85, 100, and 120 mm.*

evaluate their effects on the phase response, phase center, pattern coverage, gain, Half-Power-Beamwidths (HPBW), and polarization as compared to the equivalent FMA. The design process is complicated since several interacting parameters must be individually considered to meet the design objectives of the applications considered in this Chapter.

Figure 8 depicts the downward DMA configurations selected to perform the parametric study. A square patch is placed on three grounded substrates, L_s = 85, 100, and 120 mm. The substrates are bent at three positions in the center resulting square tops with dimensions of L_f = 10, 30, and 50 mm as shown in **Figure 8**. Three substrates have been considered with ε_r = 2.2, 4.2, and 10.

4.1 Phase response

Some GSP antenna designs have been reported in the literature [2–9, 16, 17]. However, a close scrutiny of the literature revealed that the design focused mainly on the amplitude of the radiation patterns, expressed in terms of the realized gain. On the other hand, the phase response and phase center have been reported in [23] for a flat microstrip antenna installed on a metallic choke ring. It should be noted that depending on the angle of arrival of the satellite signals, the GPS antenna far-field phase response introduces phase distortion effect, hence, the phase response must be taken into consideration to accommodate sub-centimeter positioning accuracy.

To quantitatively assess the phase performance of the DMAs, the measured phase response is matched to an ideal hemisphere using a 5° × 5° grid with equal solid angle

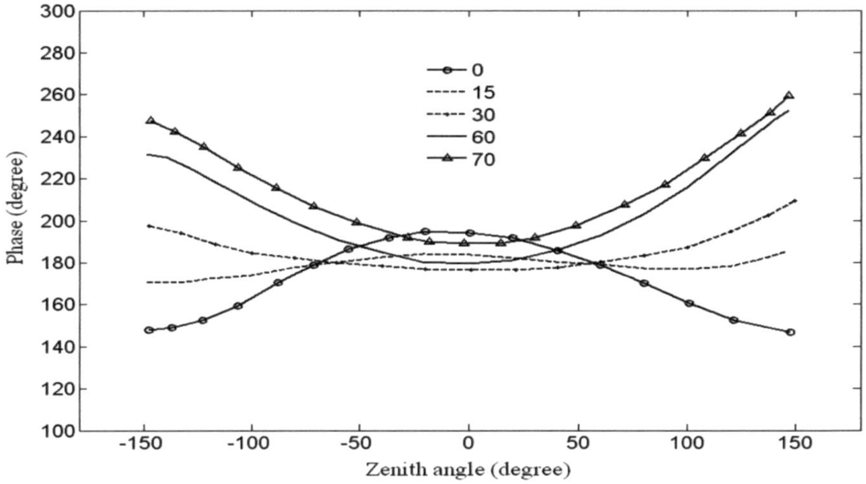

Figure 9.
Phase response for different bend angles, L_g = 100 mm, L_f = 30 mm, ε_r = 4.2.

weighting. The phase center is next obtained by adjusting the center of the hemisphere by minimizing the RMS error between the 3-D measured and the theoretical hemispherical phase response pattern. The terms, "phase residual" or "phase error" are used to denote the difference between the measured and ideal phase in the 3-D far-field region. As a Figure-of-Merit, the RMS value of the phase error to characterize the stability of the phase center of the antennas introduced in this Chapter.

Elevation cuts of the phase pattern are presented in **Figure 9** for five downward DMAs with L_g = 100 mm, L_f = 30 mm, and ε_r = 4.2. As shown in **Figure 9**, a noticeable improvement is seen in the below-horizon phase response for bend angles below 60°. As shown in **Figure 10**, it is clear that the phase response of the far-field patterns exhibits small variations above the horizon for bend angles ranging from 15° to 30°. Moreover, the RMS error in the upper hemisphere, as shown in **Figure 10**, shows little

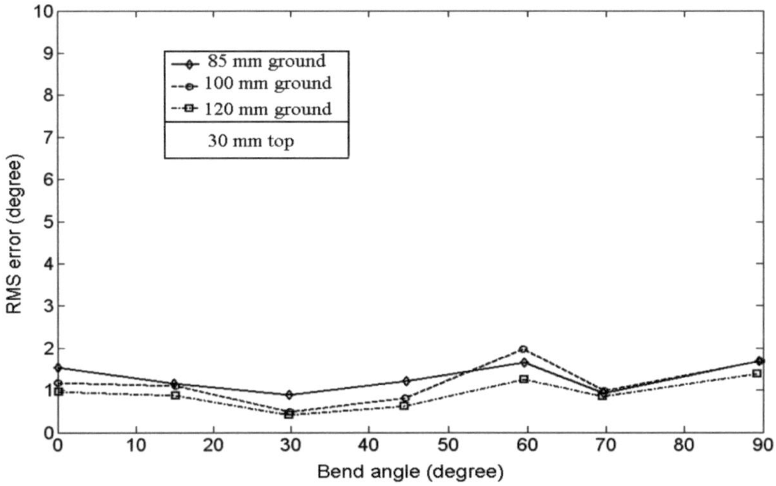

Figure 10.
RMS phase error using three ground planes, ε_r = 4.2.

variations with bend angle or ground plane size. Hence, the antennas proposed in this Chapter provide uniform phase response and stable phase center over the entire upper hemisphere when constructed with a downward scheme for bend angles between 15° and 30°. These antennas are hence suitable for precise geodetic positioning since the phase difference of the antenna outputs corresponding to different view angles is uniform which will lead to reduced positioning errors.

4.2 Radiation patterns

The measured far-field radiation patterns are measured at the resonant frequency of the GPS L1 band inside an anechoic chamber. A sequence of the normalized patterns is presented in **Figure 11** for the E_θ component in the E-plane to illustrate the range of coverage patterns that could be achieved by varying ϕ, L_f, and L_g. **Figure 11** clearly demonstrates broader beam coverage for ϕ>30° and the wide range of fields of view that can be achieved in comparison to the FMA counterparts. As expected,

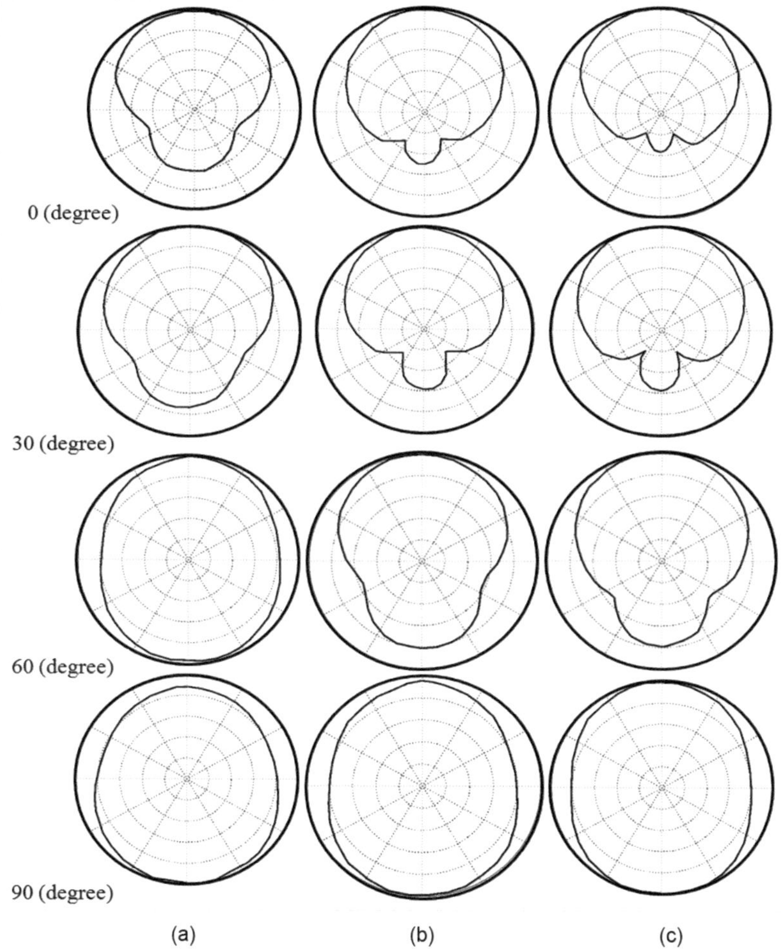

Figure 11.
Elevation patterns versus the droop angle, ϕ (a) L_g = 85 mm, (b) L_g = 100 mm, and (c) L_g = 120 mm. E_θ component, L_f = 30 mm, ε_r = 2.2, h = 1.5 mm. All patterns are normalized to 0 dB maximum, 10 dB/division.

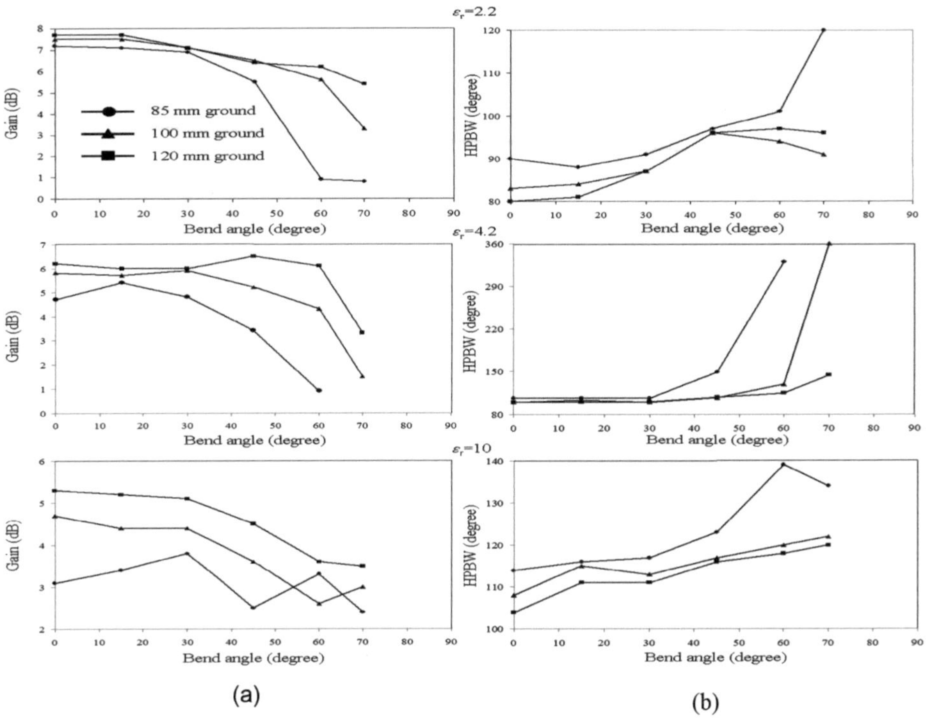

Figure 12.
Gain and HPBW for the U3DMAs versus bend angle, L_f = 10 mm, L_g = 85, 100, and 120 mm, ε_r = 2.2, 4.2, and 10 the near-horizon gain roll off is-3.7 dB.

larger bend angles on small flat tops enhance coverage due to diffraction to the extent that the main lobe appears 180° from broadside when ϕ = 90°.

Among the three flat tops considered, results are displayed for the gain and HPBW for the case of L_f = 10 mm since it provided the best performance in terms of low-elevation angle coverage. **Figure 12** shows the results for the three ground planes

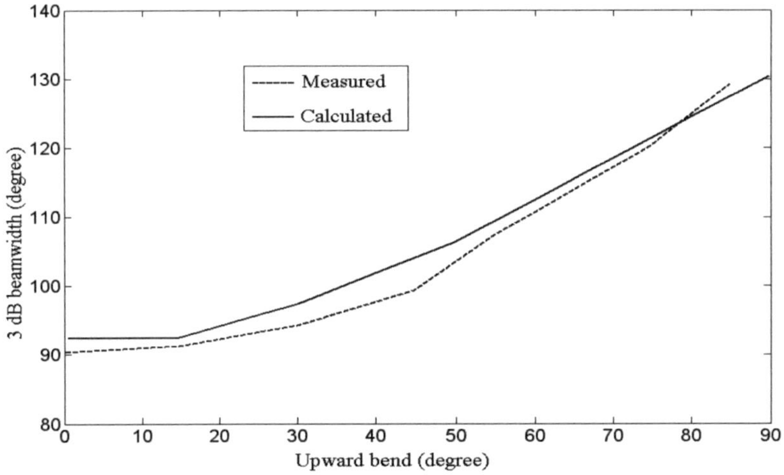

Figure 13.
Measured and simulated HPBW of an adjustable upward DMA.

and substrates under consideration. For bend angles up to 30°, the pattern coverage is found to be not significantly different from the equivalent FMA, particularly for the larger ground planes and substrates with low ε_r. It is interesting to note that an increase of 40% is observed in the HPBW with respect to the equivalent FMA for the ε_r = 4.2 substrate with L_f = 10 mm, ϕ = 45°, and L_g = 85 mm where the bore sight gain is 3.4 dB, and the near-horizon gain roll-off is −3.7 dB.

Next, the bore sight gain and HPBW are considered for the upward DMA. Simulations conducted for three bend locations: L_f = 10, 30, 50 mm; three substrates with ε_r = 2.2, 4.2, 10; and three ground planes, L_g = 85, 100, 120 mm. Compared to the equivalent flat case, the U3DMA geometries considered provided a noticeable beam broadening at bend angles greater than 60° with up to a 60% increase in the HPBW as compared to the equivalent FMA.

To verify the simulation results, a U3DMA prototype with adjustable bend plates is constructed and tested to experimentally verify the results provided by our simulation model. No additional modifications to the topology of the antenna are made as ϕ

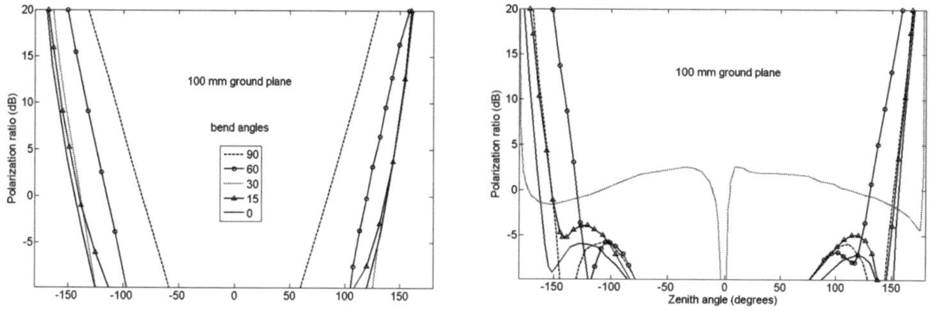

(a) ε_r = 4.2 (b) ε_r = 2.2

Figure 14.
Polarization ratio for two different substrates, L_g = 100 mm, and five downward bend angles: (a) ε_r = 4.2, (b) ε_r = 2.2.

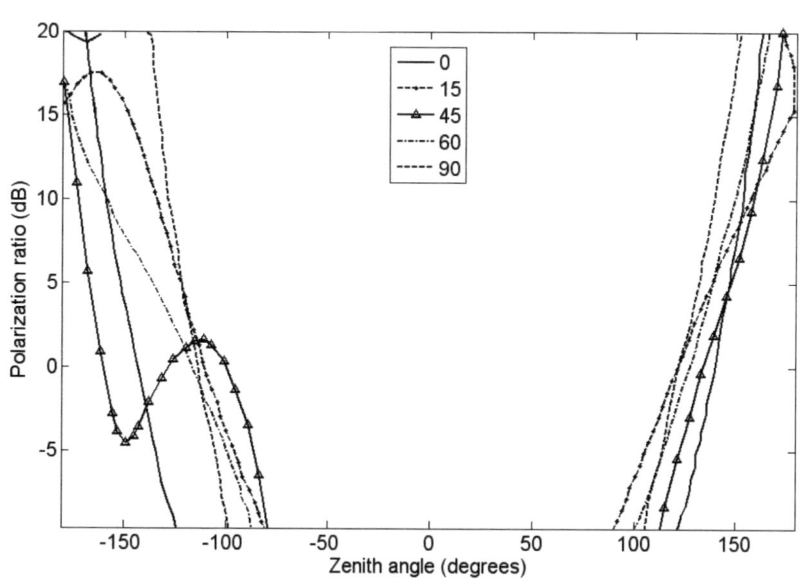

Figure 15.
Axial ratio for different upward bend angles, L_g = 100 mm, ε_r = 2.2.

changes. As can be observed from **Figure 13**, a good agreement is observed between simulated and measured HPBWs.

4.3 Circular polarization purity

In this subsection, the polarization performance of the proposed antenna structures is investigated. The polarization performance near the horizon as shown in **Figure 14** is degraded with reference to the flat microstrip structure. As is the case for traditional flat microstrip antennas, the proposed antennas are not capable of rejecting multipath signals arising from reflection, diffraction, and scattering.

Similarly, the polarization ratio for the upward structures is shown in **Figure 15**. It is clear that the AR is better near the horizon when compared to downward bending. Below the horizon, the downward bending shows better performance. It can be concluded that in both cases, drooping the antenna increases the cross-polarization level near the horizon.

5. Conclusions

In this chapter, several 3DMA which provide a uniform phase response and stable phase center are presented for bend angles ranging from 15° to 30°, which are essential to achieve sub-centimeter accuracies in GPS receivers using carrier phase measurements. The excellent phase performance demonstrated for the moderately bent structures is of paramount importance if circumstances require bending the element but without a significant change in the radiation pattern of the FMA. It is interesting to note that this improvement is accompanied by a slight reduction in the bore sight gain and the polarization purity of the corresponding FMA. The same conclusion applies to the upward drooping, except for a sharp drop in gain observed for the 100 mm ground plane, 10 mm flat top, and the $\varepsilon_r = 2.2$ substrate. The RMS phase error of the downward DMAs ranges from 0.42° to 4.3° for the three ground plane sizes, substrates, and flat portions considered for bend angles ranging from 0° to 90°. The corresponding RMS phase error for the upward DMAs ranges from 0.8° to 9.1°. For both the downward and upward DMAs, the RMS phase error decreases when the permittivity of the substrate increases and for increasing flat tops. In general, the downward DMA outperforms the upward DMA in terms of the RMS phase error performance.

On the other hand, higher bend angles allow improved GPS tracking performance for highly-dynamic marine navigation and space-borne applications. The length of the flat top, L_f and the bend angle, ϕ are found to be instrumental in the control of the radiation pattern. The upward DMA demonstrated a wider range of beam coverage compared to the FMA and downward counterparts. The half-power-beamwidth can be increased by up to 40 and 80% for the downward and upward bends, respectively, with respect to the traditional flat microstrip patch using the 10 mm flat top portion.

Finally, it is significant to stress that the design process of the 3DMA involves inevitable tradeoffs between achieving wide beam coverage, uniform phase response, gain, and polarization purity. If broad beam coverage is of precedence, some level of compromise will obviously be needed in regard to gain and multipath rejection, which requires excessive gain roll-off at low-elevation angles. Nonetheless, it turns out that the polarization purity of the proposed antenna has not significantly deteriorated since a maximum reduction of only 3 dB has been observed at the horizon in contrast to the flat microstrip structure.

Author details

Ken G. Clark[1], Jim M. Tranquilla[1] and Hussain M. Al-Rizzo[2]*

1 EMR Microwave Technology Corporation, Fredericton, Canada

2 Systems Engineering Department, Donaghey College of Science, Technology, Engineering, and Mathematics, University of Arkansas Little Rock, Little Rock, USA

*Address all correspondence to: hmalrizzo@ualr.edu

References

[1] Scire-Scappuzzo F, Makarov SN. A low-multipath wideband gps antenna with cutoff or non-cutoff corrugated ground plane. IEEE Transactions on Antennas and Propagation. 2009;**57**(11):33-46

[2] Lachapelle G, Casey M, Eaton RM, Kleusberg A, Tranquilla J, Wells D. GPS marine kinematic positioning accuracy and reliability. The Canadian Surveyor. 1987;**41**(2):143-172

[3] Altshuler EE. Hemispherical Coverage Using a Double-Folded Monopole. IEEE Transactions on Antennas and Propagation. 1996;**44**(8):1112-1119

[4] Zhang Y, Hui HT. A printed hemispherical helical antenna for GPS receivers. IEEE Microwave and Wireless Components Letters. 2005;**15**(1):10-12

[5] Feller W. Three Dimensional Microstrip Patch Antenna. US Patent Publication No. 5,200,756. April 1993

[6] Su CW, Huang SK, Lee CH. CP microstrip antenna with wide beamwidth for GPS band application. Electronics Letters. 2007;**43**(20):1062-1063

[7] Al-Rizzo HM, Clark KG, Tranquilla JM, Adada RA, Elwi TA, Rucker D. Enhanced low-angle GPS coverage using solid and annular microstrip antennas on folded and drooped ground planes. IEEE Transactions on Antennas and Propagation. 2009;**AP-57**(11): 3668-3672

[8] Tranquilla JM, Best SR. A study of the quadrifilar helix antenna for global positioning systems (GPS) applications. IEEE Transactions on Antennas and Propagation. 1990;**38**:1545-1550

[9] Shumaker PK, Ho CH, Smith KB. Printed half-wavelength quadrifilar helix antenna for gps marine applications. Electronics Letters. 1996;**32**:153-154

[10] Best SR. Distance-measurement error associated with antenna phase-center displacement in time-reference radio positioning systems. IEEE Antennas and Propagation Magazine. 2004:13-22

[11] Padros N, Ortigosa JI, Baker J, Iskander MF, Thornberg B. Comparative study of high-performance gps receiving antenna designs. IEEE Transactions on Antennas and Propagation. 1997;**45**(4):698-706

[12] Boccia L, Amendola G, Massa GD. A dual frequency microstrip patch antenna for high-precision GPS applications. IEEE Antennas and Wireless Propagation Letters. 2004;**3**:157-160

[13] Basilio LI, Williams JT, Jackson DR, Khayat MA. A comparative study of a new GPS reduced-surface-wave antenna. IEEE Antennas and Wireless Propagation Letters. 2005;**4**:233-236

[14] Zhou Y, Koulouridis S, Kiziltas G, Volakis JL. A novel 1.5" quadruple antenna for tri-band GPS applications. IEEE Antennas and Wireless Propagation Letters. 2006;**5**:224-227

[15] Basilio LI, Chen RL, Williams JT, Jackson DR. A new planar dual-band GPS antenna designed for reduced susceptibility to low-angle multipath. IEEE Transactions on Antennas and Propagation. 2007;**55**(8):2358-2366

[16] Zhou Y, Chen C-C, Volakis JL. Single-fed circularly polarized antenna element with reduced coupling for GPS arrays.

IEEE Transactions on Antennas and Propagation. 2008;**56**(5):1469-1472

[17] Yaesh I, Prielm B. Design of leveling loop for marine navigation system. IEEE Transactions on Aerospace and Electronic Systems. 1993;**29**(2):599-604

[18] Lachapelle G, Cannon ME, Lu G, Loncarevic B. Ship borne GPS attitude determination during MMST-93. IEEE Journal of Oceanic Engineering. 1996;**21**(1):100-105

[19] Lu G, Cannon ME, Lachapelle G. Attitude determination using dedicated and nondedicated multiantenna GPS sensors. IEEE Transactions on Aerospace and Electronic Systems. 1994;**30**(4):1053-1058

[20] Bull B, Oiehl J, Montenbruck O, Markgraf M. Flight performance evaluation of three GPS receivers for sounding rocket tracking. National Technical Meeting Proceeding, ION GPS, Integrating Technology, January 28-30, 2002, San Diego, California. 2002

[21] Montenbruck O, Markgraf M, Hassenpflug F. Pre-flight Assessment of a Dual Blade Antenna System for GPS Tracking of Sounding Rocket. DLR-GSOC TN 01-03. Oberpfaffenhofen: Deutsches Zentrum für Luft- und Raumfahrt; 2001

[22] Jurgens TG, Taflove A, Umashankar KR, Moore TG. Finite–difference time-domain modeling of curved surfaces. IEEE Transactions on Antennas and Propagation. 1992;**AP-40**:357-366

[23] Tranquilla JM, Carr JP, Al-Rizzo HM. Analysis of a choke ring ground plane for multipath control in global positioning system (GPS) applications. IEEE Transactions on Antennas and Propagation. 1994;**42**:905-911

Chapter 2

The Present Situation and Development for Spaceborne Synthetic Aperture Radar Antenna Arrays

Hua Li, Zhenning Li, Kaiyu Liu, Mingshan Ren and Yunkai Deng

Abstract

Synthetic aperture radar (SAR) based on satellites and or space vehicles as motion platform has the capability to work under all weather conditions, day and nights, and has become an indispensable mean for earth observation. At present, the large-scale phased array antenna, such as the active microstrip-phased array antenna of SIR-C imaging radar and the waveguide slot-phased array antenna of Radarsat-1 imaging radar, is one of the core components of spaceborne SAR and determines the system performance to a large extent. Although the traditional antenna array scheme has been widely used in existing spaceborne SAR systems due to strong beam control ability, sufficient failure redundancy backup, mature design method and so on, it still has drawbacks, for example, large volume, high weight and manufacturing cost, and low energy utilization rate, which restrict the further improvement of performance and are burdens on the research of next-generation high-performance spaceborne SAR system. With the development of electric and electronic techniques, forthcoming SAR array-phased array antennas will make breakthroughs in antenna architectures, concepts, technologies and modes, for instance, periodic reflector array antennas and metamaterial array antennas. This chapter focuses on the present and forthcoming development of spaceborne SAR antenna arrays.

Keywords: spaceborne synthetic aperture radar, antenna arrays, beamforming, metamaterials, high-resolution wide-swath

1. Introduction

In 2018, an English named Ian Wilson claimed that he had found the MH370 passenger plane which had been missing for four and a half years on Google maps and it was located in a dense forest in Cambodia. The news caused a sensation on the global Internet. Changguang Satellite Technology Co., Ltd. immediately maneuvered all 10 satellites to observe this area one after another. Unfortunately, the first photos were disappointing. The authenticity of the rumor could not be confirmed because the suspected crash site of MH370 was covered by clouds. This dramatic event reminded people of synthetic aperture radar (SAR), because it can make up for the shortcomings of optical earth observation by the capability to pass through all kinds of obstacles such as clouds, rain, snow, fog, sand, dust, and so on and to work under all weather conditions, day and nights [1–3].

Spaceborne SAR equipped on low/medium orbit satellites or other orbital platforms such as space shuttle, the upper stage of the launch vehicle, and so on is a kind of advanced earth observing and imaging system [4–6]. It has played an irreplaceable role in the fields of global military reconnaissance, environmental remote sensing, natural disaster monitoring, and planetary exploration and gradually become a research hotspot in the field of earth observation since the United States launched the first SAR satellite SeaSat in 1978 [7, 8]. Many countries have successively carried out spaceborne SAR technology research and formulated corresponding development plans of the SAR satellite system. In the 21st century, many countries have successively deployed their own spaceborne SAR satellite systems and realized the upgrading of SAR satellites. For example, European Space Agency (ESA) launched Sentinel-1 to replace ENVISAT [9]. In recent 10 years, spaceborne SAR has made great progress in hardware structure, imaging theory, system performance, and application fields. It is worth noting that only the TecSAR system adopts the reflector antenna and the rest are phased array antenna [5]. In fact, an array antenna is the main configuration scheme for most modern high-performance spaceborne SAR, because it is preferred to obtain better performance within the limited mission budget [10, 11]. This chapter focuses on the development trend of array antenna technology of spaceborne SAR and discusses the research status in the future from the aspects of system composition, antenna performance, planar phased array antenna, paraboloid array antenna, and metasurface array antenna.

2. Spaceborne SAR antenna subsystem technical features

Spaceborne SAR system usually consists of operation loads and function loads. The former mainly serves the spaceborne platform operation, such as the attitude and orbit control subsystem, telemetry control subsystem, environmental control subsystem, power subsystem, and so on. The function loads directly related to SAR imaging processing are usually composed of the following subsystems:

- SAR antenna subsystem

It can radiate the radar pulse signal generated by the front-end subsystem and receive the corresponding radar echo signal. The antenna needs to be able to meet the strict requirements of beam scanning and shaping for spaceborne SAR.

- Radio-frequency signal generating and processing subsystem

According to the task requirements, it generates the modulated pulse signal transmitted by the SAR antenna and down converts as well as samplings the received original radar echo signal. After analog-to-digital conversion, the original imaging data are obtained, which is handed over to the rear terminal system for imaging processing.

- Signal processing and imaging subsystem

It completes the imaging and image information compression processing of the original data in a high-performance data processor.

- Data handling and transmission subsystem

It stores the image information as well as other relevant data and transmits them to the ground receiving station.

Figure 1 shows the photo of Canada's Radarsat-2 satellite and deployable 12 m × 3 m lightweight SAR membrane antenna. The long strip structure covered with gold foil below is C-band SAR Antenna.

The SAR antenna subsystem which occupies a considerable amount of mass as well as volume and consumes the most power during operation is the most difficult, time-consuming, and costly part, due to the stringent requirements on antenna performance. In general, spaceborne SAR mainly has the following technical features:

- Wide observation strip and large observation area

The commonly observed swath width can reach 40 km or more and the length of continuous imaging area can reach thousands of kilometers. Through antenna beam scanning and satellite platform maneuver, the width of the observable area can reach 500–600 km, including more than 10 common observation strips [7].

- Far slant range and high signal transmission path loss

For example, for a spaceborne SAR system with a center frequency of 9.6 GHz, an orbit height of 600 km, and an antenna look angle of 25°–45°, the slant range

Figure 1.
Canada's Radarsat-2 satellite and deployable SAR Membrane Antenna.

between the antenna to the observation area is 670–890 km and the two-way path loss is 168–170 dB.

- Long revisit cycle period

The satellite platform runs along the preplanned orbit and can only observe the same area once in a revisit cycle. Even if the orbit planning is reasonable and the antenna beam has good adjustment ability, the typical revisit cycle of spaceborne SAR still ranges from 2 to 3 days to more than 10 days [7, 8]. If forced orbit change is required to shorten the revisit period, additional satellite fuel will be consumed and the system life and mission duration will be shortened.

- Serious ambiguity signal interference

Spaceborne SAR system faces inherent ambiguity signal problems, including range ambiguity and azimuth ambiguity. The intensity and distribution of the ambiguous signal are affected by the width of the imaging area, pulse repeat frequency, pulse width, pulse bandwidth, look angle, backscattering coefficient, and other parameters [12]. Most of the above parameters restrict each other and are difficult to control independently. In order to reduce the ambiguity signal ratio to meet the system index requirements, it is necessary to decrease the SAR antenna pattern sidelobe to the source direction of the ambiguity signal.

- Various working modes

Currently, spaceborne SAR systems in operation or under development can generally work in Stripmap mode, ScanSAR mode, Spotlight mode, etc. Each mode has different requirements for antenna beam characteristics [13]. **Figure 2** displays the corresponding beam diagrams of five spaceborne SAR operating modes given from left to right, including huge region ScanSAR mode, wide region ScanSAR, himage Stripmap mode, polarimetric Stripmap mode, and spotlight mode. Generally

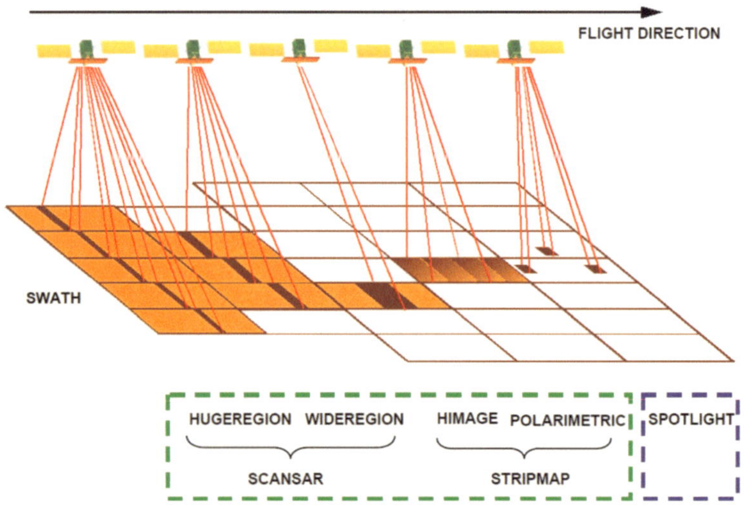

Figure 2.
Various working modes for spaceborne SAR.

speaking, according to different task requirements, the scanning beam ranges from several degrees to tens of degrees in the range direction while the azimuth beam scans within several degrees.

- Wide frequency band

In order to meet different mission requirements, the commonly used spaceborne SAR operating frequency band spans from P-band to X-band. Electromagnetic waves with different frequency bands have different penetrating abilities. Generally speaking, the higher the frequency, the wider the bandwidth and the higher the resolution, but the weaker the penetration ability to vegetation and other covers. Ku, K, and Ka band spaceborne SAR satellites with higher resolution belong to the next generation system under research have not been widely used at present.

- Full polarimetric signal processing

In addition, with the development of SAR target polarimetric information processing technology, more and more spaceborne SAR systems have the ability of full polarimetric signal processing.

According to the above technical features of the spaceborne SAR system, a series of strict performance requirements of array antenna is determined as follows:

- Narrow beam and high gain

In order to compensate for the extremely high path loss, the spaceborne SAR antenna needs to have a very high gain. Taking the X-band TerraSAR-X system launched in 2007 as an example [8], its SAR antenna in **Figure 3** size is 4.8×0.75 m and the antenna normal gain exceeds 46 dBi. At the same time, in order to suppress the ambiguous signal and improve the imaging quality, the spaceborne SAR antenna has strict restrictions on the maximum width of the main beam for range and azimuth. In addition, COSMO-Skymed's, which was launched in 2007 and works in X-band [14], antenna size is 5.7×1.4 m. The average width of the range beam is 1.5° and the average width of the azimuth beam is only 0.3°.

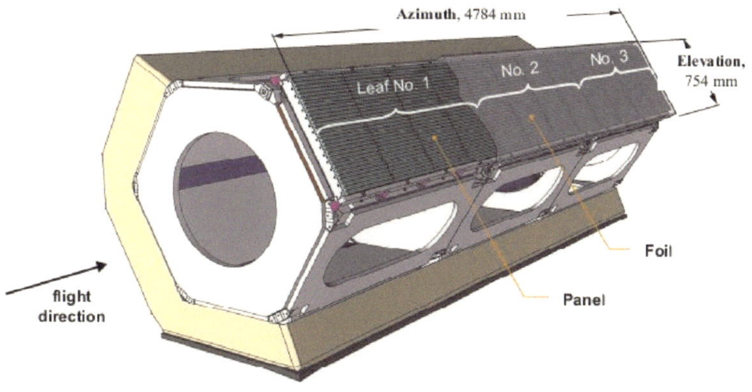

Figure 3.
Diagram of antenna structure for TerraSAR-X.

- Beamforming and sidelobe suppression

The sidelobe level in the source direction of the ambiguity signal needs to be especially suppressed to ensure that the ambiguity signal ratio meets the system index requirements. Compared with TX pattern, which takes the maximization of effective isotropic radiation power (EIPR) as the design guide and does not particularly consider sidelobe suppression, the RX pattern in **Figure 4** has to carry out a wide range of sidelobe suppression in the main range direction of ambiguity signal source. At the same time, in order to ensure that the echo signal level in the imaging area does not exceed a certain range to ensure the imaging quality, the main beam lobe needs to be shaped and adjusted, which makes the amplitude of the pattern point to the far end for the observation area appropriately higher than that pointing to the near end, so as to compensate for echo signal attenuation caused by larger path loss and smaller ground friction angle [15].

- High EIRP of TX antenna and high G/T value of RX antenna

In order to ensure that the spaceborne SAR system can successfully complete the imaging mission, the actually received signal power and signal-to-noise ratio of the data acquisition device need to meet the minimum requirements, which involves two factors that, first, the EIPR of the TX antenna should be high enough, that is, in addition to the high gain, the actual transmitting power of the antenna should also reach a high level. On the other hand, in order to achieve a sufficient signal-to-noise ratio of the received signal, the RX antenna is required to have a sufficiently high G/T value.

- Full polarimetric mode

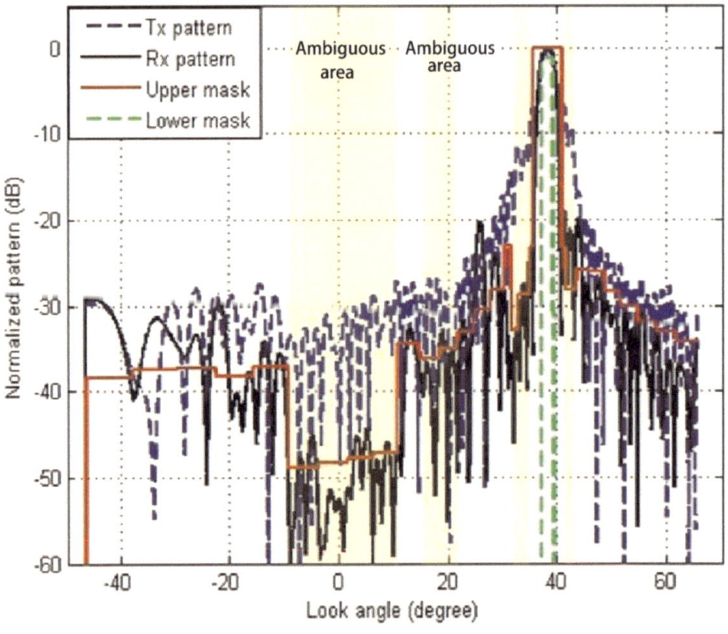

Figure 4.
Range beam pattern of typical SAR Antenna.

Early spaceborne SAR systems, such as SeaSat in 1978 and SIR-C in 1981, used a single polarimetric antenna. In recent years, as more and more spaceborne SAR systems have the ability of full polarimetric signal processing, there is an urgent need for the full polarimetric performance of SAR Antenna. At present, there are two main types of spaceborne SAR antenna: microstrip patch array antenna and waveguide slot array antenna. The former, represented by COSMO-Skymed in **Figure 5** [7] and Radarsat-2 [6], is usually a double-layer microstrip patch array based on lightweight dielectric materials [16, 17]. It has the advantages of light structure weight, dual-polarization, flexible arrangement of elements and subarrays, etc., but it has a slightly larger loss compared with the other scheme. The latter is represented by TerraSAR-X [18] and sentinel-1 in **Figure 6** [19, 20]. The ridged waveguide wide-edge slot antenna [21] and the reduced waveguide narrow-edge slot antenna made of carbon fiber reinforced plastic [22, 23] are alternately arranged to form a dual polarimetric array antenna, which has low loss and high radiation efficiency, but it is slightly heavy, difficult to design as well as process and difficult to adjust the position of array elements and subarrays.

- Limited range scanning angle and small azimuth scanning angle

The orbit altitude of spaceborne SAR platform is usually between 500 and 800 km. Due to the limitations of earth curvature and range resolution, the antenna beam in the range direction, that is, the direction on the elevation plane, is usually within the range of 15°–50° starting from the connecting line between the satellite and the nadir on the earth's surface, with a span of 35°. Limited by the Doppler bandwidth of the processed echo signal, the beam coverage and scanning range in the azimuth, that is, horizontal plane, are usually within ±2° [5, 7, 17]. Compared with the phased array

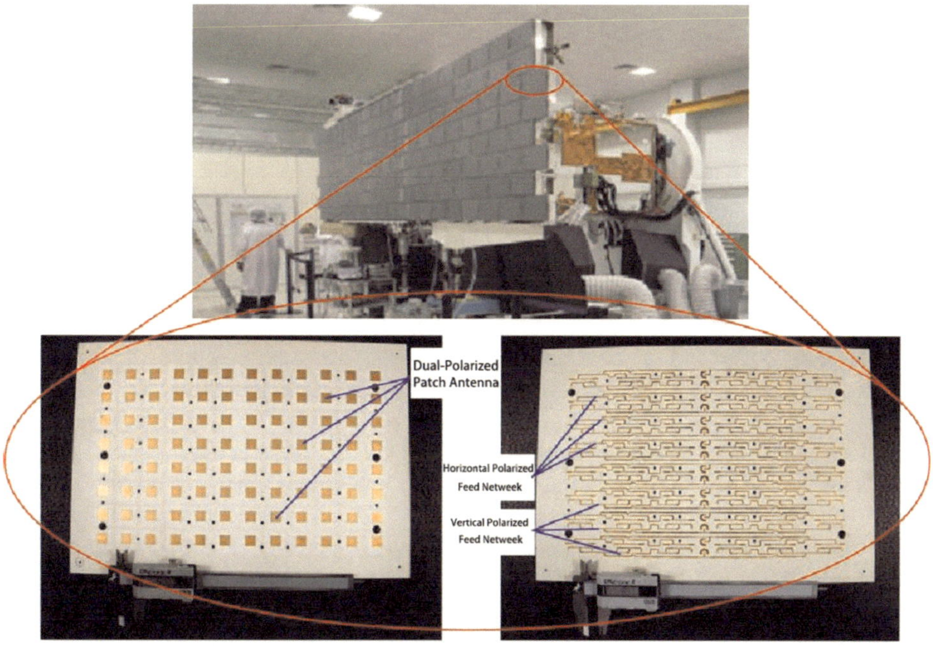

Figure 5.
Photo of microstrip patch array SAR antenna of COSMO-Skymed.

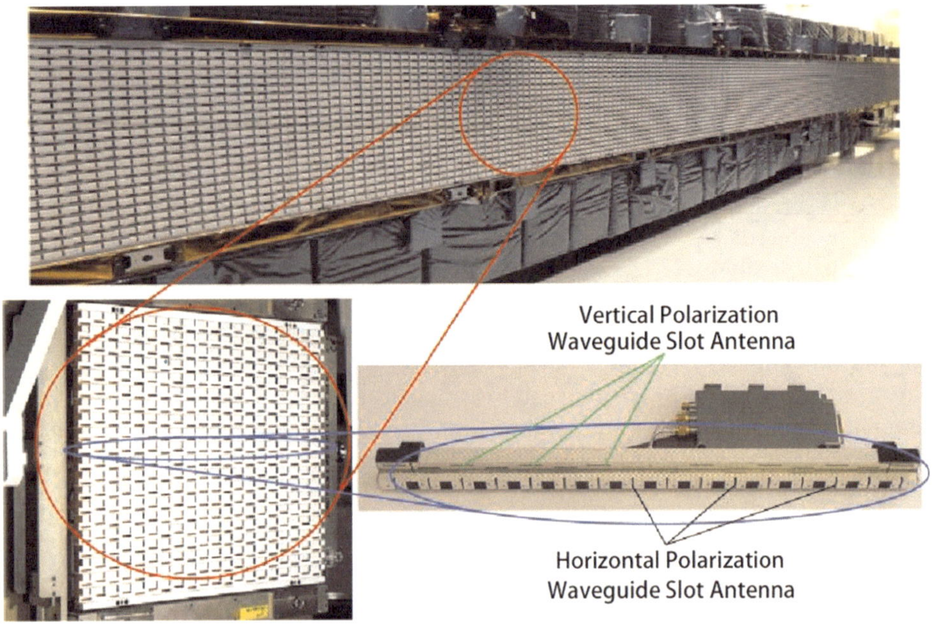

Figure 6.
Photo of waveguide slot array SAR antenna of Sentinel-1.

antenna of other applications, such as warning/fire control radar, the scanning range of spaceborne SAR antenna is limited in range and very small in the azimuth.

3. Technical features and development of spaceborne SAR planar phased array antenna

Planar phased array antenna has been widely used in spaceborne SAR systems and its superior performance has been fully demonstrated and verified. **Table 1** summarizes the features and technical parameters of some spaceborne SAR antennas using planar phased array antennas.

Planar phased array antenna uses a medium-scale line array as the basic unit to form a full-scale area array. Since the scanning angle of the azimuth beam is very small, according to the basic principle of the phased array antenna, the size of the array element in this direction or the phase center spacing of the adjacent array elements can be expanded to achieve the purpose of reducing the number of control channels without reducing the number of control channels. Grating lobes appear when the beam is scanned. The scanning angle of the distance beam is relatively large, and it is necessary to ensure that the phase center spacing of adjacent units is less than 0.8 times the wavelength of the center frequency to ensure that the sidelobe performance does not deteriorate excessively during beam scanning. Therefore, the current mainstream spaceborne SAR phased array antennas usually use 10–20 minimum radiating elements (microstrip patches or waveguide slots) to form a linear array, and the excitation of each element inside the linear array cannot be controlled independently. After that, the linear array is used as the basic unit of the full-size antenna to control the amplitude and phase of each linear array. Generally speaking, there are 12–

Working waveband	Name	Center frequency (GHz)	Pulse bandwidth (MHz)	Relative bandwidth (%)	Range of view angle (°)	Antenna element type	Antenna size (m^2)	Launch time
P	Seasat	1.275	19	1.5	20–26	microstrip patch	10.7 × 2.2	1978
	SIR-A	1.275	6	0.47	47–53	microstrip patch	9.4 × 2.16	1981
	JERS-1	1.275	15	1.2	32–38	microstrip patch	11.9 × 2.2	1992
	PALSAR	1.27	28	2.2	9.9–50.8	microstrip patch	—	2006
S	NovaSAR-S	3.2	200	6.25	16–34	microstrip patch	3 × 1	2013
C	ERS-1/2	5.3	15.55	3	20.1–25.9	waveguide slot	11 × 1.3	1991/1995
	RADARSAT-1	5.3	30	0.57	20–58	microstrip patch	15 × 1.5	1995
	RADARSAT-2	5.405	100	1.85	20–50	microstrip patch	15 × 1.37	2007
	ENVISAT	5.331	124	2.32	15–45	microstrip patch	10 × 1.3	2002
	SENTINEL-1	5.405	100	1.85	20–45	waveguide slot	12.3 × 0.9	2014
X	COSMO-SkyMed	9.6	400	4.17	19–49	microstrip patch	5.7 × 1.4	2007
	TerraSAR-X	9.65	300	3.1	25–55	waveguide slot	4.8 × 0.7	2007
	Paz	9.65	300	3.1	25–55	waveguide slot	4.8 × 0.7	2017

Table 1.
The features and technical parameters of spaceborne SAR antennas.

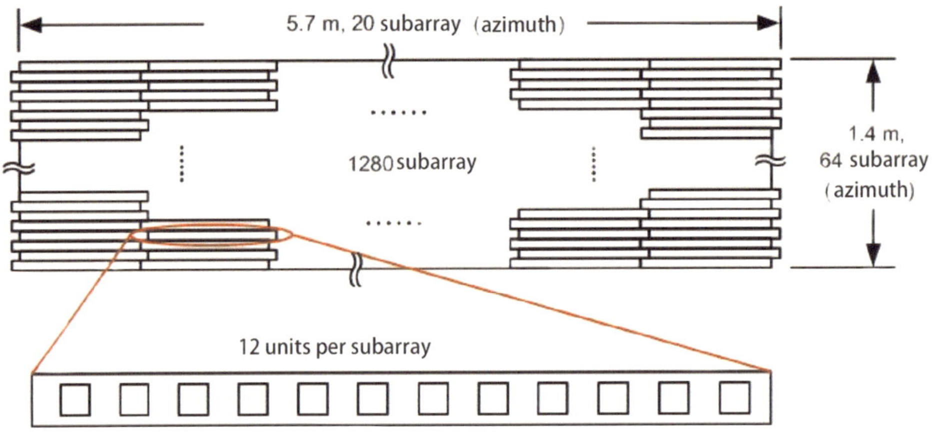

Figure 7.
Conceptual schematic diagram of COSMO-Skymed phased array antenna.

20 line arrays along the azimuth direction, and 30–64 line arrays along the distance direction. The entire array can have 300 to more than 1000 line arrays and corresponding amplitude and phase control channels, and more than 10,000 actual minimum radiation unit (microstrip patch or waveguide slot). **Figure 7** shows a conceptual schematic diagram of the COSMO-Skymed phased array antenna structure. The antenna consists of 64 (elevation plane) × 20 (azimuth plane) a total of 1280 line arrays, each line array contains 12 basic radiating elements. All line arrays are arranged in the horizontal (azimuth plane) direction. Note that in order to improve the grating lobe performance, a staggered arrangement is used between adjacent linear arrays on the pitch plane, but in principle, it is still a planar linear array.

In addition, its relative bandwidth is not large. At present, the relative bandwidth of the transmitted pulses of spaceborne SAR systems using phased array antennas usually does not exceed 5%. For example, COSMO-Skymed has a working center frequency of 9.6 GHz, a maximum bandwidth of 400 MHz and a relative bandwidth of 4.2%. Sentinel-1 has a working center frequency of 5.405 GHz, a maximum bandwidth of 100 MHz and a relative bandwidth of 1.85% [4, 19]. In fact, this is also an inevitable limitation brought by the use of medium-scale linear arrays as the minimum amplitude and phase control unit system. For example, the bandwidth of a waveguide slot antenna with more than 15 slots is difficult to exceed 6% [24]. The Nova-SAR with a smaller antenna size (3 × 6 linear arrays, each with 24 patch elements) has the widest working bandwidth among spaceborne SAR systems equipped with microstrip patch phased array antennas and its working center frequency is 3.2 GHz. The maximum bandwidth is 200 MHz and the relative bandwidth is 6.25% [3].

However, the traditional spaceborne SAR periodic array phased array antenna still has the following shortcomings:

- Expensive

The total price of a large number of high-performance RF active devices is so considerable that some spaceborne SAR users with urgent needs and limited budgets can only use other antenna solutions [5].

- Excessive total system weight and widely distributed mass

This aspect occupies a large proportion of the launch payload, but also makes the SAR satellites have a large moment of inertia and a high fuel consumption rate for maneuvering and attitude control during in-orbit operation [25–27].

- High Power consumption

When working at full power, the phased array antenna will consume most of the power supply of the satellite platform, and the peak power consumption far exceeds the power provided by the solar panels on the satellite. For this reason, SAR satellites are usually equipped with large-capacity batteries. The issue of power consumption has become one of the main factors limiting the working time of continuous imaging of spaceborne SAR. Taking the COSMO-Skymed system, which began to operate in orbit in 2007, as an example, the peak power required for SAR imaging observation in spotlight mode is 17.3 kW, and most of the energy is consumed by the phased array antenna (especially the transmitter); while the solar panels onboard the satellite can only provide a maximum of 4.5 kW (at the initial stage of the mission) to 3.5 kW (at the end of the mission). To make up for this gap, a lithium-ion battery with a capacity of 336 Ah is onboard the satellite. The battery weighs 136 kg, and the mass of the entire satellite at launch is only about 1700 kg (including fuel and propellant). Even with this power supply configuration, the COSMO-Skymed only lasts about 10 s of continuous imaging when operating in the most power-hungry spotlight mode, and only about 10 s when operating in the lower-power stripe mode. Minutes limit the information acquisition capability of spaceborne SAR to a large extent. One of the important reasons for this problem is that the transmit-receive power transition efficiency (TRPE) of the current mainstream spaceborne SAR periodic array phased array antenna is relatively low.

- Redundant system performance

As far as the general spaceborne SAR system needs to scan and shape the antenna beam, the periodic array phased array antenna is actually in a state of excess performance. Specifically, phased array antennas that achieve the smallest antenna aperture area of spaceborne SAR are of considerable size, usually containing more than 300, or even more than 1000 independent transceiver control channels. For a phased array antenna of this size, even if up to 10% of the transceiver units fail, the excitation of the remaining units can still be adjusted to roughly maintain the main radiation performance indicators of the antenna, such as beam deflection angle, main lobe width and the sidelobe level in the ambiguous area, which is equivalent to a certain amount of redundancy in the antenna performance. In the traditional concept of spaceborne SAR engineering practice, it is generally believed that the performance of some antenna unit components will gradually degrade or fail during the operation of the mission. At this time, it is necessary to release the redundant performance of the array to ensure that the SAR antenna function can still be properly used at the end of the mission cycle [28, 29]. Function properly. However, in recent years, with the development of advanced solid-state active RF circuit technologies such as Monolithic Microwave Integrated Circuit (MMIC), Low-Temperature Co-fired Ceramics (LTCC), and spaceborne SAR systems. With the improvement of orbital operation management experience, the reliability of phased array antenna units and radio frequency components used in spaceborne SAR is increasing day by day. The number

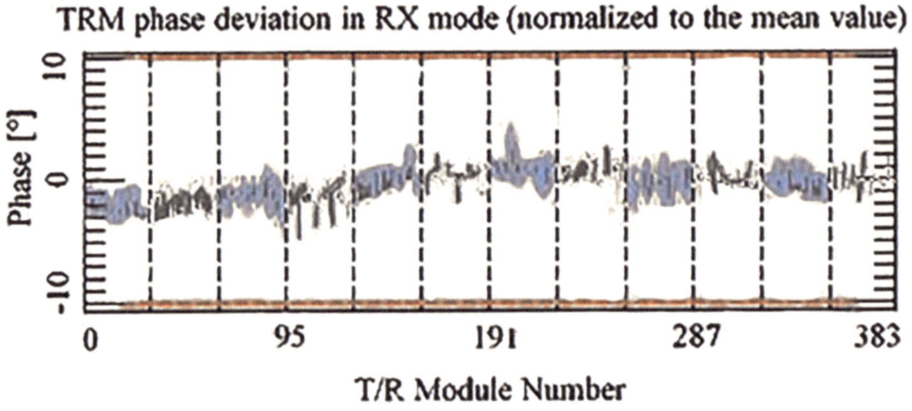

Figure 8.
Phase error test results of T/R components for TerraSAR-X (2016).

of failed units during the full mission cycle is often lower than expected during mission planning [30–33]. For example, the CosmoSky-Med (No. 1 Star) and TerraSAR-X systems [8], they were launched by ESA in 2007 and 2008 and then started operating, respectively. The original mission periods of the two satellites are only until 2012 [34, 35]. However, as of the end of 2016, it is still operating reliably in orbit, and the phased array SAR antenna carried is still in good condition. **Figure 8** shows the phase error test results of all 384 T/R components of TerraSAR-X in the eighth year (2016) in orbit. The phase errors of all channels are far less than the tolerance range of 10° (significant difference). Significant difference. It can be seen that the scale of the traditional spaceborne SAR phased array antenna has a certain compressible space under the condition of maintaining the system performance generally unchanged.

The deficiencies of large phased array antennas in the above four aspects have restricted the development and application of spaceborne SAR systems to a certain extent. Studies have shown that the above problems are expected to be improved when applying Uniform Amplitude/Equal Amplitude or Quantized Amplitude aperiodic array phased array antennas in spaceborne SAR systems [36–39]. The spacing between the elements of an Aperiodic Phased Array Antenna is usually not uniform [36, 38, 40, 41]. This is the most significant difference from the periodic array phased array. The number of aperiodic and periodic array elements of the same aperture is not necessarily the same. Generally speaking, the number of practical aperiodic array elements is less than or equal to that of the periodic array of the same aperture. When the number of aperiodic array elements is smaller than that of periodic arrays with the same aperture, it can also be called a thinned array or a sparse array. For sparse arrays, the element spacing is an integer multiple of a certain greatest common divisor; while for sparse arrays, the spacings of array elements are randomly distributed and have no greatest common divisor.

A series of studies have discussed the advantages of aperiodic array phased array antennas in reducing the number of active devices and improving energy utilization efficiency when applied to spaceborne SAR [36, 39, 42, 43] and synchronous satellite communication systems [37, 38, 44–52]. For these two types of applications, there are three things in common: (1) high-gain beams; (2) strict sidelobe control; and (3) beam scanning requirements over a relatively small angular range. In order to achieve high-gain beams, the array antenna must have a larger aperture and a larger number of elements; strict sidelobe control means that the excitation amplitude of the antenna array must be significantly tapered; the beam scanning requirement of a smaller angle makes the restrictions on the maximum spacing of the antenna elements in the medium are looser. The above three characteristics make it possible to design aperiodic array antennas excited by equal amplitude [36] or quantized amplitude [38, 53] on the basis of tapered amplitude excitation periodic array antennas. Such aperiodic array antennas naturally have the advantages of reducing the number of active devices and improving energy utilization efficiency.

A widely studied and applied aperiodic array antenna design is the density tapering (Density Tapering) constant amplitude excitation aperiodic array antenna [39, 44, 54, 55]. Compared with the Amplitude Tapering periodic array antenna used as a design reference, a well-designed density-tapered equal-amplitude excited aperiodic array antenna has roughly equivalent gain and sidelobe performance, and can be used as an active transmit antenna. All high-power amplifiers are made to work at the saturation operating point of the highest efficiency so that the DC-RF Power Transition Efficiency (DC-RF Power Transition Efficiency) is higher than that of the traditional periodic array antenna whose excitation amplitude is tapered. **Figure 9** presents the results of a typical constant-amplitude excitation aperiodic array design [39]. Among them, the upper left is the comparison of the aperiodic array element position, excitation amplitude, and the reference excitation amplitude used as a reference to taper the periodic array; the upper right is the pattern of the aperiodic and periodic array antennas, especially the sidelobe performance comparison, it can be seen that

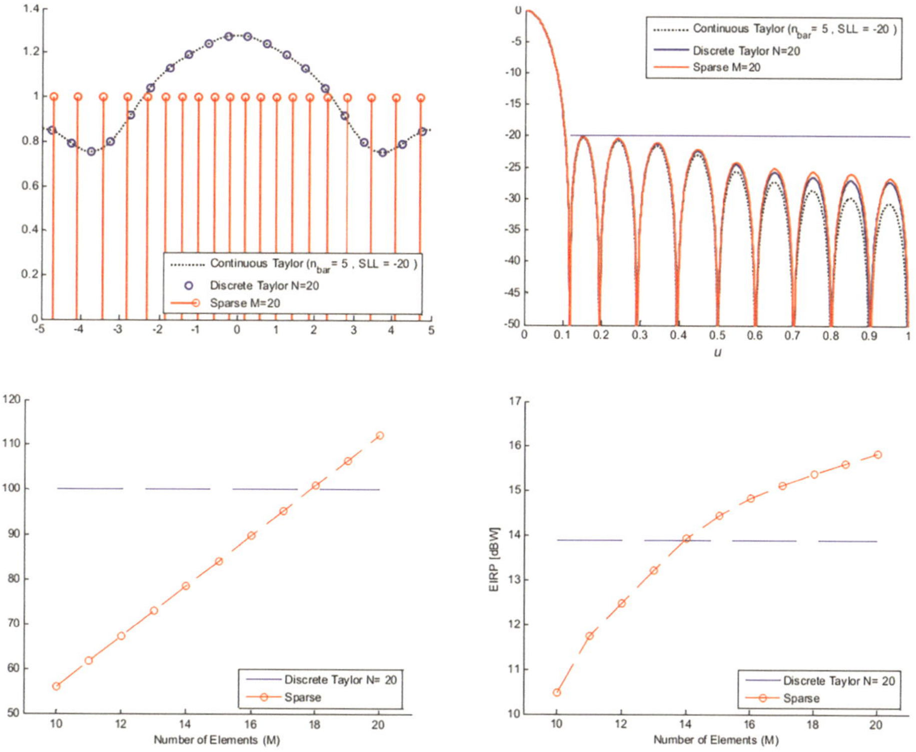

Figure 9.
Example of density tapered aperiodic array antenna design results.

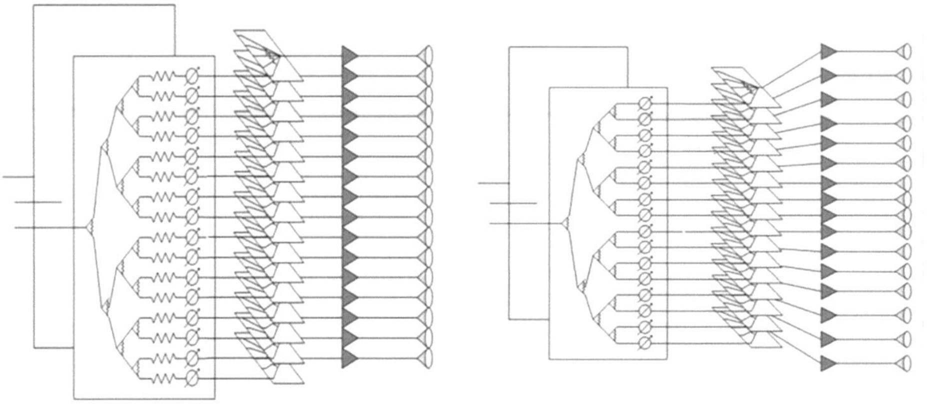

Figure 10.
Diagram of amplitude taper periodic array phased array and equal-amplitude aperiodic array phased array structure.

the two are basically equivalent; the lower left is the variation of the DC power consumption of the aperiodic array with the number of elements, and compared with the case of the 20-element periodic array, the lower right is the variation of the Equivalent Isotropic Radiation Power (EIRP) of the aperiodic array with the number of elements and is compared with the case of the 20-element periodic array. Compared with the 20-element periodic array, it can be seen that when the number of elements is between 14 and 18, the aperiodic array consumes less DC power than the 20-element periodic array and obtains a higher EIRP.

Compared with periodic array antennas, aperiodic array antennas have very different characteristics. As shown in the right of **Figure 10** [39], firstly, the analysis of the array characteristics can no longer be simplified to the processing of one element in the periodic boundary; secondly, the characteristics of each element in the aperiodic array are quite different; finally, the processing and Test work is also much more difficult and complex than periodic arrays. Despite the above difficulties, the research on aperiodic array antennas has still received extensive attention. In addition to the aforementioned advantages of reducing the number of radiation units or control units and improving the DC-RF conversion efficiency, the aperiodic array actually increases the degree of freedom in the design of the array antenna, so it is expected to obtain a more ideal array design result. Characteristics are also one of the reasons why it is studied.

In recent years, a group of European research institutions funded by ESA has carried out research on spaceborne aperiodic phased array antenna experimental systems mainly serving satellite communication systems. In 2010, a team of researchers from Naples University, Università di Cassino and Space Engineering S.P. A. reported the results of their preliminary development of an experimental system for satellite communications aperiodic array antennas. The designed aperiodic array antenna is composed of various aperture units, as shown in the left in **Figure 11**. The test results prove that the antenna design has high aperture efficiency and global beam coverage. The same group of researchers also completed the design of a multi-beam aperiodic dielectric lens antenna based on a similar theoretical design, as shown on the right in **Figure 11**. The experimental test results show that the performance of the designed aperiodic array antenna is basically the same as that of the periodic array antenna used for comparison, and the number of control devices is significantly reduced [37].

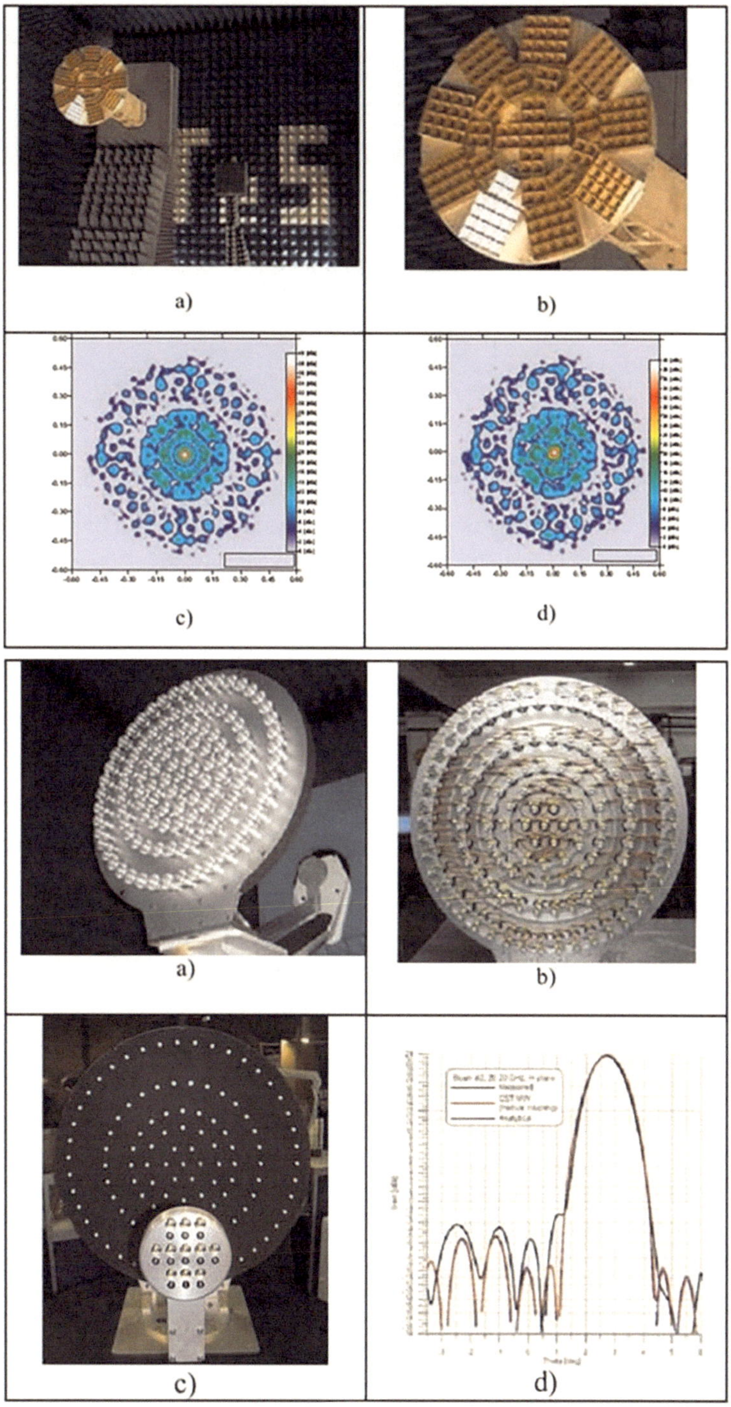

Figure 11.
ESA satellite communication aperiodic array antenna experimental system hybrid unit direct radiation array and test results (left), aperiodic reflection array and test results (right).

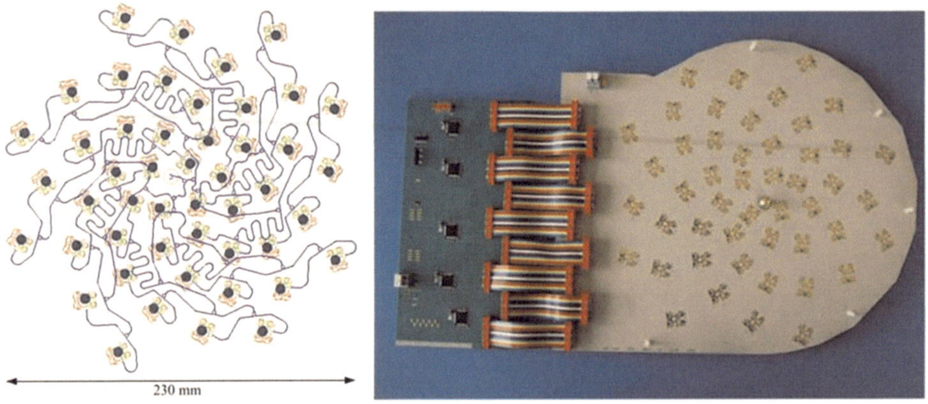

Figure 12.
Aperiodic sparse array phased array antenna experimental system for satellite communication on the ground.

In 2014, Maria Carolina Viganó of ViaSat Antenna Systems in Switzerland and others completed the research and development of the ground-side aperiodic sparse array phased array antenna experimental system for satellite communication with the special support of ESA [45]. The system works in the Ka-band and adopts 1-bit quantization phase control, and its design is based on a novel sunflower arrangement, as shown in **Figure 12** [56]. The left picture shows the array topology and the distribution of the feeding network, and the right picture shows the finished experimental sample. The test results show that the antenna has reached the basic design index and has good development potential.

The application of aperiodic array phased array antenna in spaceborne SAR systems began in the 21st century. ESA and the National Aeronautics and Space Administration (NASA) have funded a number of research institutions to carry out research on Spaceborne SAR sparse array or aperiodic array phased array antenna, and have made some research progress in array synthesis technology, aperiodic array imaging technology, and hardware implementation scheme.

In terms of aperiodic array synthesis technology for Spaceborne SAR requirements, ESA and Cristian luison and Stefano selleri of the University of Florence in Italy published an important research paper on the application of aperiodic array phased array antenna to spaceborne SAR in 2012 [36]. This paper presents a practical method for Spaceborne SAR aperiodic array synthesis, which is divided into three steps: firstly, according to the task requirements of spaceborne SAR, the synthesis of a single excitation amplitude distribution pattern based on the traditional periodic array is completed; secondly, according to the results of the amplitude distribution of periodic array, the aperiodic sparse array is obtained by amplitude density taper transformation; finally, the element position and excitation phase of the sparse array are optimized to make the final design results meet the beam performance requirements. As shown in **Figure 13**, the upper part of the figure shows the three-step design process and the schematic diagram of array element distribution obtained in each step. Starting from the initial 64-element periodic array, a 48-element linear array with constant amplitude excitation is obtained after the second step of thinning, and then the final aperiodic array unit position and excitation phase distribution are obtained through the third step of array element position optimization. The lower part of **Figure 13** shows the patterns generated by the designed aperiodic array to meet the beam requirements of two different spaceborne SAR. Both patterns are generated by

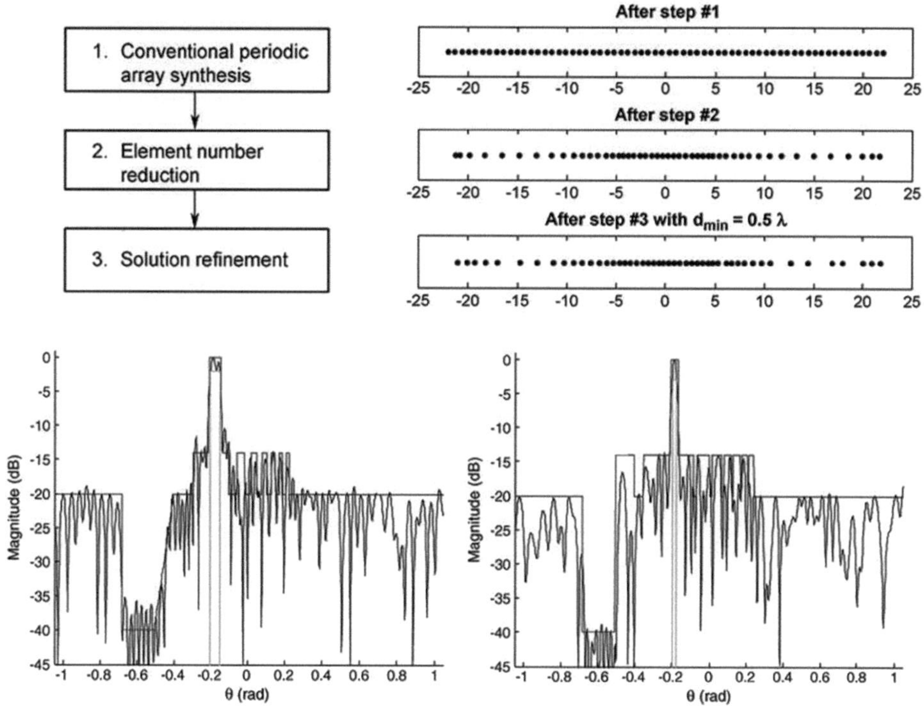

Figure 13.
Results of spaceborne SAR aperiodic array antenna.

the same aperiodic array with equal amplitude excitation, and only the excitation phase distribution is different.

Although it can significantly improve the system performance in theory, the engineering realization of aperiodic array antenna is very difficult, and the development of related technologies is not mature. Therefore, from the perspective of ensuring system reliability and avoiding technical risks, the conditions for applying aperiodic array phased array antennas in spaceborne SAR systems with high R&D costs and a small number are not yet mature.

4. Technical features and development of spaceborne SAR parabolic phased array antenna

TanDEM-L is a spaceborne distributed SAR satellite plan proposed by Deutsches Zentrum für Luft-und Raumfahrt (DLR). The plan consists of two L-band spaceborne SAR systems. The schematic diagram of the concept is shown in **Figure 14**. The program was developed to monitor dynamic changes on the Earth's surface such as forest biomass, millimeter-scale surface deformation, polar ice, and soil moisture. In the TanDEM-L plan, the inversion of vegetation parameters, stereo imaging, and surface deformation monitoring technology are realized through polarization interference, multi-baseline, and heavy orbit technology, which are helpful for a better understanding of the earth system, risk analysis, disaster management, and environmental monitoring are of great significance (**Figure 14**) [57].

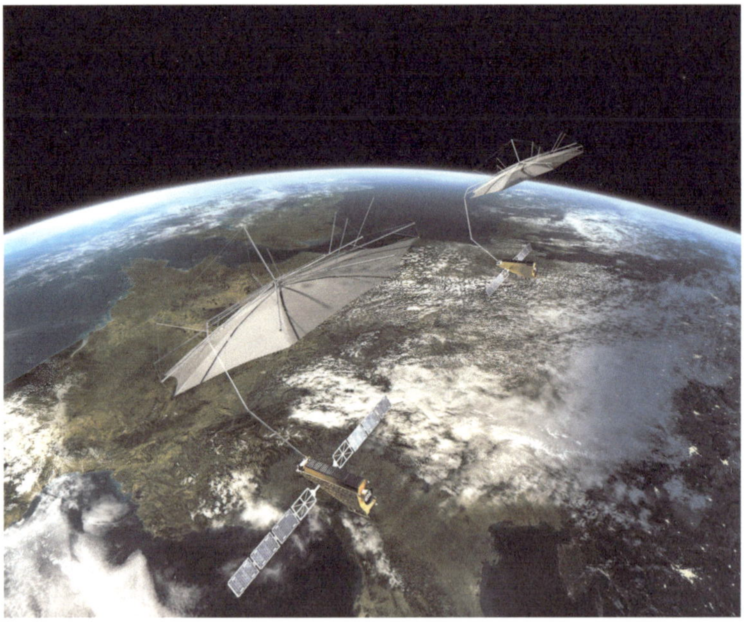

Figure 14.
Preview of TanDEM-L Plan.

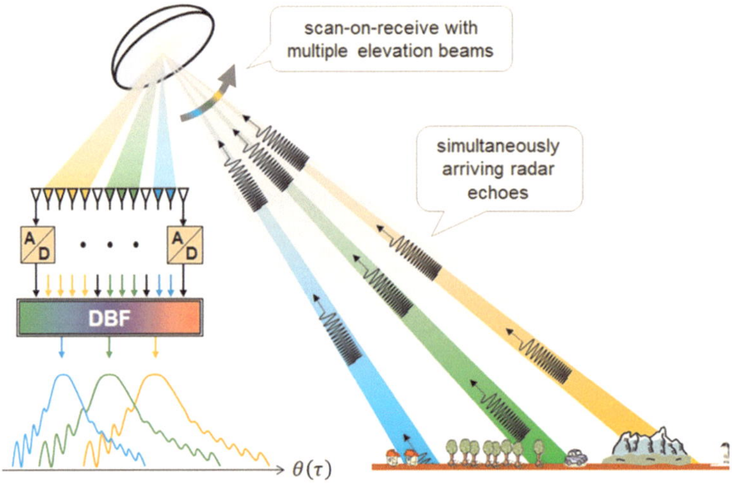

Figure 15.
Example of Elevation multibeam reflector SAR system.

Technologies such as parabolic reflector antenna, DBF technology (as shown in **Figure 15**), and variable PRF technology are used in the TanDEM-L satellite program. Through the combination of these technologies, TanDEM-L plans to achieve 3 m resolution, 350 km wide Earth observation, and the ability to realize a dynamic observation of the world in 6 days. Among these technologies, the combination of reflector antenna and DBF technology is used to achieve high gain acceptance through distance scanning and improve the system signal-to-noise ratio; variable PRF technology changes the position of the blind spot to achieve continuous imaging of the scene [58].

A major feature of the TanDEM-L program is the use of a combination of a deployable reflector antenna and a two-dimensional digital feed array (**Figure 16**). The left image is a technical solution proposed by DLR researchers to combine a reflector antenna and a two-dimensional digital feed array. In this scheme, the diameter of the reflector antenna is 15 m, and the feed source is a digital array of 36 (elevation) × 3 (azimuth) and aligned horizontally. The scheme achieves a wide range and high gain in the distance direction by using the scanning processing concept of "Scan-On-Receive". Each of the pitch array elements consists of dual-polarized microstrip patches, and the signal from each polarization port is routed through the TR assembly and digitized in the corresponding DBF unit [58].

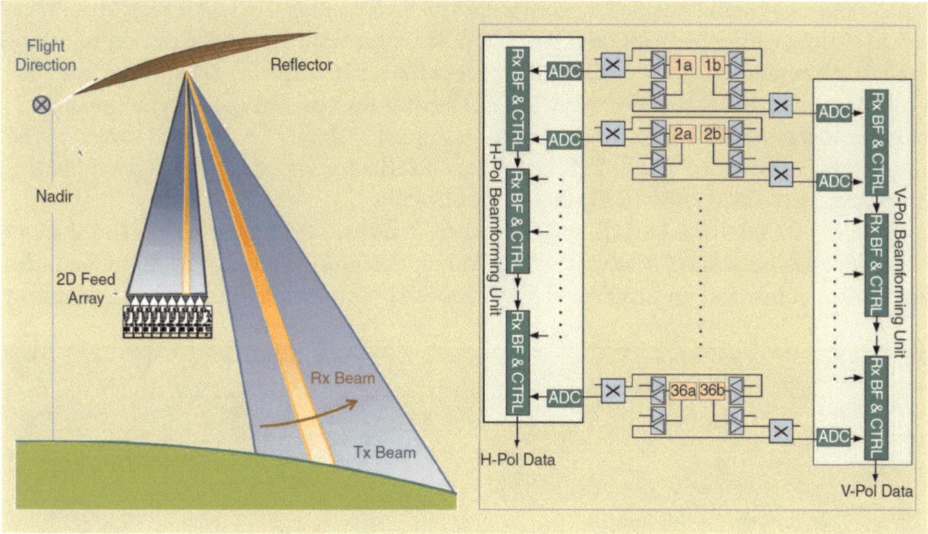

Figure 16.
Reflector and feed array of TanDEM-L.

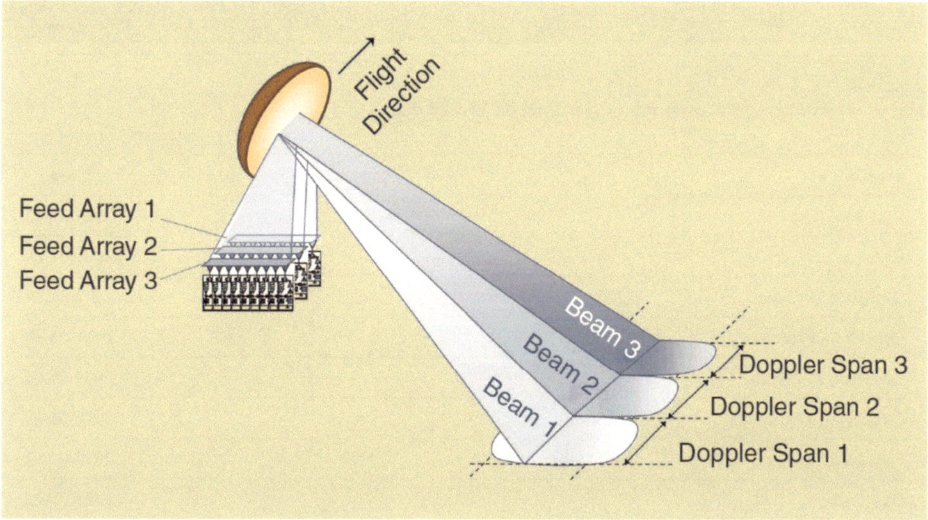

Figure 17.
Three azimuthal feed arrays antenna of TanDEM-L.

This scheme achieves high resolution in azimuth by adding the feed array and azimuth beamforming technology in **Figure 16** along the azimuth. As shown in **Figure 17**, each feed array corresponds to an azimuth channel transformed to the ground, and the overall azimuth spectrum is recovered by beamforming technology.

5. Technical features and development of spaceborne SAR plane reflector metasurface phased array antenna

The Capella X-SAR satellite constellation plan is a constellation plan proposed by Capella Space, which consists of 36 spaceborne SAR systems. 36 SAR satellites form 12 orbital planes, each operating in a 500 km polar orbit with an orbital period of 90 min, providing average imaging revisit time of less than 1 h. Capella's 01 star Denali was launched as a test star in December 2018. The star uses an "origami-type" planar reflector metasurface antenna (as shown in **Figure 18** left). The launch time of subsequent stars is shown in **Table 2**. It is a meshed reflector antenna with a peripheral frame structure (as shown in **Figure 18** right) [59].

Capella's 01 satellite, Denali, was launched as a test satellite, which is the size of a backpack and has a mass of about 40 kg. When designing Denali's SAR payload, the researchers combined an origami-shaped antenna with efficient electronics. Among

Figure 18.
Plane reflector metasurface array antenna of Capella X-SAR plan.

Satellite	Date
Capella 2 (Sequoia)	2020.08.31
Capella 3 (Whitney 1)	2021.01.24
Capella 4 (Whitney 2)	2021.01.24
Capella 5 (Whitney 3)	2021.06.30
Capella 6 (Whitney 4)	2021.05.15
Capella 7 (Whitney 5)	2022.01.13
Capella 8 (Whitney 6)	2022.01.13
Capella 9	2022(plan)

Table 2.
Capella X-SAR plan.

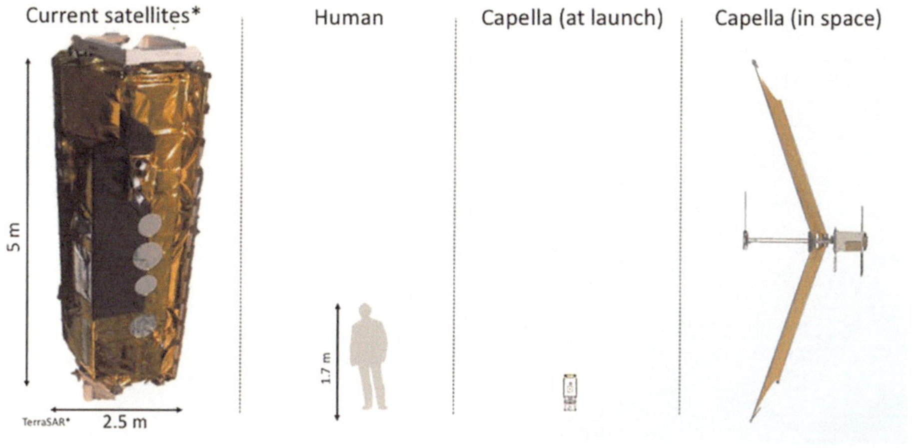

Figure 19.
Example of array antenna of Capella satellite.

them, the antenna is made of flexible material, and the area is about 8 m^2 after deployment. Columns 3 and 4 in **Figure 19** are the dimensions of the Denali satellite before and after deployment [60].

6. Conclusions

A large-scale array antenna is one of the core components of high-performance spaceborne SAR, which determines the performance of spaceborne SAR to a great extent. The traditional spaceborne SAR phased array antenna adopts the periodic array implementation scheme, which has the advantages of strong beam control ability, sufficient failure redundancy backup, mature design method, and so on. It has been widely used in the existing spaceborne SAR system. However, the phased array antenna based on the periodic array scheme has the disadvantages of large size, heavy weight, high manufacturing cost, and low energy utilization, which restricts the further improvement of the performance of the spaceborne SAR system and the research and development of the next generation of high-performance spaceborne SAR system. Using aperiodic array antenna, parabolic array antenna and hypersurface array antenna have the possibility to overcome the above shortcomings. Based on the technical characteristics and development trend of spaceborne SAR array antenna, this article reviews the theoretical analysis, simulation results, and experimental verification.

Acknowledgements

The author would like to thank Y. Wen and J. Wang for their tremendous help and technical discussions.

Author details

Hua Li*, Zhenning Li, Kaiyu Liu, Mingshan Ren and Yunkai Deng
Department of Space Microwave Remote Sensing System, Aerospace Information Research Institute, Chinese Academy of Sciences, Beijing, China

*Address all correspondence to: fiskli@foxmail.com

References

[1] Elachi C, Bicknell T, Jordan RL, Chilin W. Spaceborne synthetic-aperture imaging radars: Applications, techniques, and technology. Proceedings of the IEEE. 1982;**70**:1174-1209. DOI: 10.1109/PROC.1982.12448

[2] L band Mission – Seasat 1 [EB/OL]. 2014. Available from: https://directory.eoportal.org/web/eoportal/satellite-missions/s/seasat

[3] S band NovaSAR-S [EB/OL]. 2015. Available from: https://directory.eoportal.org/web/eoportal/satellite-missions/n/novasar-s

[4] C band Copernicus: Sentinel-1 [EB/OL]. 2014. Available from: https://directory.eoportal.org/web/eoportal/satellite-missions/c-missions/copernicus-sentinel-1

[5] X band TecSAR [EB/OL]. 2008. Available from: https://directory.eoportal.org/web/eoportal/satellite-missions/t/tecsar

[6] C band RADARSAT-2 [EB/OL]. 2007. Available from: https://directory.eoportal.org/web/eoportal/satellite-missions/r/radarsat-2

[7] COSMO-SkyMed, eoPortal Directory [EB/OL]. 2007. Available from: https://directory.eoportal.org/web/eoportal/satellite-missions/c-missions/cosmo-skymed

[8] X band Terra-SAR[EB/OL]. 2007. Available from: https://directory.eoportal.org/web/eoportal/satellite-missions/t/terrasar-x

[9] Berger M, Moreno J, Johannessen JA, Levelt PF, Hanssen RF. ESA's sentinel missions in support of Earth system science. Remote Sensing of Environment. 2012;**120**:84-90. DOI: 10.1016/j.rse.2011.07.023

[10] SIR-A[EB/OL]. 1981. Available from: https://directory.eoportal.org/web/eoportal/satellite-missions/s/sir-a

[11] Rosenqvist A, Shimada M, Watanabe M. L band ALOS PALSAR: Technical outline and mission concepts. In: 4th International Symposium on Retrieval of Bio- and Geophysical Parameters from SAR Data for Land Applications; Innsbruck, Austria, 16-19 November 2004. pp. 1-7

[12] Li FK, Johnson W. Ambiguities in spaceborne synthetic aperture radar systems. IEEE Transactions on Aerospace and Electronic Systems. 1983;**19**:389-397. DOI: 10.1109/TAES.1983.309319

[13] X band COSMO-SkyMed System Description and User Guide [EB/OL]. 2007. Available from: http://www.e-geos.it/products/pdf/csk-user_guide.pdf

[14] Torre A, P C. X band COSMO-SkyMed: The advanced SAR instrument. In: 5th International Conference on Recent Advances in Space Technologies (RAST '11); 9–11 June 2011. Istanbul, New York: IEEE; 2011. pp. 865-868

[15] Ka M, Kononov AA. Effect of look angle on the accuracy performance of fixed-baseline interferometric SAR. IEEE Geoscience and Remote Sensing Letters. 2007;**4**:65-69. DOI: 10.1109/LGRS.2006.885858

[16] Onwujekwe O, Uguru N, Russo G, Etiaba E, Mbachu C, Mirzoev T, et al. Role and use of evidence in policymaking: an analysis of case studies from the health sector in Nigeria. Health Research Policy and Systems. 2015;**13**:46. DOI: 10.1186/s12961-015-0049-0

[17] Capece P, Borgarelli L, Di Lazzaro M, Di Marcantonio U, Torre A. Cosmo Skymed active phased array SAR instrument. In: Radar Conference (RADAR '08); 26–30 May 2008. Rome, New York: IEEE; 2008. pp. 1-4

[18] Grafmuller B, Herschlein A, Fischer C. The TerraSAR-X antenna system. In: Radar Conference (RADAR '05); 9–12 May 2005. Arlington VA. New York: IEEE; 2005. pp. 222-225

[19] Ostergaard A, Snoeij P, Traver IN, Ludwig M, Rostan F, Croci R. C-band SAR for the GMES Sentinel-1 mission. In: 8th Europena Radar Conference (EURAD'11); 12–14 October 2011; Manchester. New York: IEEE; 2011. pp. 234-240

[20] Schoenfeld U, Braudach H. Electrical architecture of the SENTINEL-1 SAR antenna subsystem. In: 7th European Conference on Synthetic Aperture Radar (EUSAR '08); 2–5 June 2008; Friedrichshafen. New York: IEEE; 2008. pp. 1-4

[21] Zhang HT, Wang W, Zhang ZH. Ridged waveguide slot antenna array with low cross-polarization. In: International Conference on Microwave and Millimeter Wave Technology (ICMMT '12); 5–8 May 2012; Shenzhen. New York: IEEE. pp. 1-3

[22] Mehdipour A, Sebak AR, Truemen CW. Conductive carbon fiber composite materials for antenna and microwave applications. In: 29th National Radio Science Conference (NRSC '12); 10–12 April 2012; Cairo. New York: IEEE. pp. 1-8

[23] Wagner R, Braun HM. A slotted waveguide array antenna from carbon fibre reinforced plastics for the European space SAR. Acta Astronautica. 1980;**8**: 273-282. DOI: 10.1016/0094-5765(81) 90036-9

[24] Zahn R, Kirscht M, Weidmann K. Modular Radar Core for airborne and space applications. IEEE International Geoscience and Remote Sensing Symposium. 2010;**2010**:677-680. DOI: 10.1109/IGARSS.2010.5651829

[25] Moussessian A, Delcastillo L, Bach V, Grando M, Quijano U, Smith P, et al. Large aperture, scanning, L-Band SAR. In: NASA Earth Science Technology Conference, June 21, 2011, Pasadena, California. pp. 1-4. DOI: 10.1049/ joe.2019.0602

[26] Leipold M, Runge H, Sickinger C. Large Sar Membrane Antennas With Lightweight Deployable Booms. In: 28th ESA Antenna Workshop on Space Antenna Systems and Technologies. 2005. pp. 1-8

[27] Wang HJ, Fan B, Yi M, Guan FL, Guang L, Xue C, et al. Inflatable Antenna for Space-Borne Microwave Remote Sensing. IEEE Antennas and Propagation Magazine. 2012;**54**:58-70. DOI: 10.1109/ MAP.2012.6348118

[28] Di Lorenzo P, Barbarossa S, Borgarelli L. Optimal Beamforming for Range-Doppler Ambiguity Minimization in Squinted SAR. IEEE Transactions on Aerospace and Electronic Systems. 2013; **49**:277-293. DOI: 10.1109/ TAES.2013.6404103

[29] Zeng XN, He F, Zhang YS, Dong Z. Analysis and compensation of spaceborne SAR antenna array deformation. Journal of National University of Defense Technology. 2012; **34**:158-163. DOI: 10.3969/j. issn.1001-2486.2012.03.031

[30] Wolff I. From antennas to microwave systems-LTCC as an

integration technology for space applications. In: 3rd European Conference on Antennas and Propagation (EuCAP '09); 23–27 March 2009; Berlin. New York: IEEE; 2009. pp. 3-8

[31] Florian C, Traverso PA, Feudale M, Filicori F. A C-band GaAs-pHEMT MMIC low phase noise VCO for space applications using a new cyclostationary nonlinear noise model. In: IEEE MTT-S International Microwave Symposium (IMS '10); 23–28 May 2010; Anaheim CA. New York: IEEE; 2010. pp. 284-287

[32] Cao MY, Zhang K, Chen YH, Zhang JC. High-efficiency S-band harmonic tuning GaN amplifier. Chinese Physics B. 2014;**23**:37305. DOI: 10.1088/1674-1056/23/3/037305

[33] Attema EPW. The Active Microwave Instrument On-Board the ERS-1 Satellite. Proceedings of the IEEE. 2002; **79**:761-799. DOI: 10.1109/5.90158

[34] Buckreuss S, Steinbrecher U, Schattler B. The TerraSAR-X Mission Status. In: Asian and Pacific Conference on Synthetic Aperture Radar (APSAR '15); 1–4 September 2015; Singapore. New York: IEEE; 2015. pp. 357-361

[35] Valerio G, Barbara B, Pasquale S, Perrera A, Pepe P, Inversi P, et al. CSK mission status and experimentation results. In: 11th European Conference on SAR (EUSAR '16); 6–9 June 2016; Hamburg. New York: IEEE; 2016. pp. 1-3

[36] Luison C, Landini A, Angeletti P, Toso G, Valle P, Capece P, et al. Aperiodic Arrays for Spaceborne SAR Applications. IEEE Transactions on Antennas and Propagation. 2012;**60**: 2285-2294. DOI: 10.1109/TAP.2012.2189714

[37] Catalani A, Russo L, Bucci O M, T. Isernia: "Sparse Arrays for Satellite Communications: From Optimal Design to Realization". Proceedings of the 32nd ESA Antenna Workshop on Antennas for Space Applications. 2010. p. 5-8. DOI: 10.1109/APS.2012.6348916

[38] Angeletti P, Toso G. Aperiodic arrays for space applications: A combined amplitude/density synthesis approach. In: 3rd European Conference on Antennas and Propagation (EuCAP '09); 23–27 March 2009; Berlin. New York: IEEE; 2009. pp. 2026-2030

[39] Toso G, Angeletti P, Mangenot C. A comparison of density and amplitude tapering for transmit active arrays. In: 3rd European Conference on Antennas and Propagation (EuCAP '09); 23-27 March 2009; Berlin. New York: IEEE; 2009. pp. 840-843

[40] Ishimaru A. Theory of unequally-spaced arrays. IRE Transactions on Antennas and Propagation. 1962;**10**: 691-702. DOI: 10.1109/TAP.1962.1137952

[41] Pierro V, Galdi V, Castaldi G, Pinto IM, Felsen LB. Radiation properties of planar antenna arrays based on certain categories of aperiodic tilings. IEEE Transactions on Antennas and Propagation. 2005;**53**:635-644. DOI: 10.1109/TAP.2004.841287

[42] Lu J, Guo Y. Novel aperiodic phased array with reduced number of active chains for space-borne SAR. In: 5th Asia-Pacific Conference on Synthetic Aperture Radar (APSAR '15); 1-4 September 2015; Singapore. New York: IEEE. pp. 20-23

[43] Lu J, Guo Y, Yang H. Sparse phased array antenna for space-borne SAR. In: International Symposium on Antennas and Propagation (ISAP '15); 19–24 July

2015; Vancouver. New York: IEEE; 2015. pp. 2461-2462

[44] Lu J, Guo Y. Compact Planar Sparse Array Antenna with Optimum Element Dimensions for SATCOM Ground Terminals. International Journal of Antennas and Propagation. 2015:806981. DOI: 10.1155/2015/806981

[45] Vigano MC, Llorens Del Rio D, Bongard F, Bongard F, Vaccaro S. Sparse Array Antenna for Ku-Band Mobile Terminals Using 1 Bit Phase Controls. IEEE Transactions on Antennas and Propagation. 2014;**62**:1723-1730. DOI: 10.1109/TAP.2014.2301439

[46] Skobelev SP. One more look at the reduction of the number of controlled elements in limited-scan phased array antennas. In: 8th European Conference on Antennas and Propagation (EuCAP '14); 6-11 April 2014; The Hague. New York: IEEE; 2014. pp. 3625-3628

[47] Angeletti P, Toso G. Array antennas with jointly optimized elements positions and dimensions part I: Linear arrays. IEEE Transactions on Antennas and Propagation. 2014;**62**: 1619-1626. DOI: 10.1109/ TAP.2013.2281602

[48] Angeletti P, Toso G, Ruggerini G. Array Antennas with Jointly Optimized Elements Positions and Dimensions Part II: Planar Circular Arrays. IEEE Transactions on Antennas and Propagation. 2014;**62**:1627-1639. DOI: 10.1109/TAP.2013.2281519

[49] Bucci OM, Isernia T, Perna S, Pinchera D. Isophoric sparse arrays ensuring global coverage in satellite communications. IEEE Transactions on Antennas and Propagation. 2014;**62**: 1607-1618. DOI: 10.1109/ TAP.2013.2287901

[50] Johansson S. Design Study into Sparse Array Antenna Demonstrators for Future Satellite Systems[D]. Gothenburg: Ghalmers University of Technology; 2014

[51] Catalani A, Russo L, Isernia T, Toso G, Angeletti P. Ka-Band active sparse arrays for SATCOM applications. In: International Symposium on Antennas and Propagation (ISAP '12); 8-14 July 2012; Chicago. New York: IEEE; 2012. pp. 1-2

[52] Toso G, Mangenot C, Roederer AG. Sparse and thinned arrays for multiple beam satellite applications. In: 2nd European Conference on Antennas and Propagation (EuCAP '07); 11-16 November 2007; Edinburgh. New York: IEEE; 2008. pp. 1-4

[53] Lu J, Yang H, Wen G, Li D. An aperiodic phased array antenna for space-borne Synthetic Aperture Radar. In: Loughborough Antennas & Propagation Conference (LAPC '16); 14-15 November 2016; Loughborough. New York: IEEE; 2017. pp. 1-5

[54] Caratelli D, Viganó MC. Analytical synthesis technique for linear uniform-amplitude sparse arrays. Radio Science. 2011;**46**:1-6. DOI: 10.1029/ 2010RS004522

[55] Bucci OM, D'Urso M, Isernia T, Angeletti P, Toso G. Deterministic synthesis of uniform amplitude sparse arrays via new density taper techniques. IEEE Transactions on Antennas and Propagation. 2010;**58**:1949-1958. DOI: 10.1109/TAP.2010.2046831

[56] Viganó MC, Toso G, Caille G, Cyril M, Lager I. Sunflower array antenna with adjustable density taper. International Journal of Antennas and Propagation. 2009;**2009**:1-10. DOI: 10.1155/2009/624035

[57] Younis M et al. Tandem-L instrument design and SAR performance overview. IEEE Geoscience and Remote Sensing Symposium. 2014;**2014**:88-91. DOI: 10.1109/IGARSS.2014.6946362

[58] Klenk P, Reimann J, Schwerdt M. Performance aspects of large-deployable reflector antennas based on surface deformations simulated for tandem-L. In: 2019 13th European Conference on Antennas and Propagation (EuCAP). 2019. pp. 1-5

[59] Stringham C et al. The Capella X-band SAR constellation for rapid imaging. In: IGARSS 2019 – 2019 IEEE International Geoscience and Remote Sensing Symposium. 2019. pp. 9248-9251. DOI: 10.1109/IGARSS.2019.8900410

[60] Castelletti D, Farquharson G, Stringham C, Duersch M and Eddy D."Capella Space First Operational SAR Satellite". 2021 IEEE International Geoscience andRemote Sensing Symposium IGARSS. 2021. p. 1483-1486. DOI:10.1109/IGARSS47720.2021.9554100

Chapter 3

Radiation Patterns from Thinned and Unthinned Linear Arrays with Different Spacings Using ML Algorithms

M. Laxmi Prasanna Rani, Moturi Satyanarayana and Narsupalli Shanmukha Rao

Abstract

Antenna array thinning is the tuning of same antenna elements with uniform spacing or periodic antenna array to generate the desired amplitude density across the aperture area. Large antenna arrays are difficult to build and have increased fabrication cost. The process of eliminating the radiating elements from the array would be desirable if arrays performance is not significantly degraded. One method of achieving this goal is array thinning by systematically removing elements without change in the performance. This chapter presents and explores machine learning and its applications in the design of antenna array. This chapter also gives the characteristics of machine learning, deep learning, different learning algorithms and its usage in the design of an antenna array with thinning. This chapter presents the performance of an array with and without thinning and the radiation characteristics are observed for both the cases with different spacings. The major advantage of the present work is the reduction of number of elements to achieve better and specified Radiation patterns.

Keywords: Thinning, Spacing, linear array, machine learning algorithm

1. Introduction

Artificial Intelligence (AI) is an approach to make a robot, or a product to think how smartly humans can think. It is the study of how the human brain thinks, learns, decides, and works, when it attempts to solve the problems. And finally this study outputs intelligent software systems. By using algorithms and large amount of data sets, iterative processing permits AI software to learn automatically. Artificial Intelligence (AI) and Machine Learning(ML) are similar and can be used interchangeably. **Figure 1** shows the relationship between ML, AI, and Deep Learning. ML is a huge subset of AI. In fact, ML can be used as an approach to achieve applications of AI. These ML algorithms are very powerful in optimization, but the performance of these algorithms depends upon the size of the data collected [1–3]. Therefore, ML is often

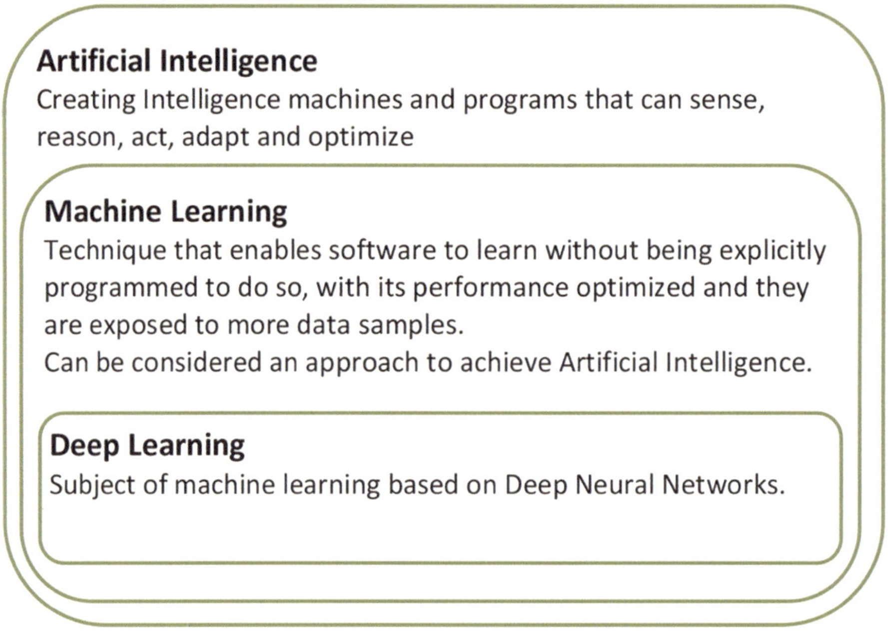

Figure 1.
Relation among artificial intelligence, deep and machine learning.

accompanying with data statistics and data analysis. Deep Neural Networks (DNN) are neural networks which contain more than one hidden layer. DNN consist of layers of interconnected nodes/neurons where each and every node/neuron produces a nonlinear function of its input given. All these algorithms are considered as the types of machine learning algorithms [4].

Single-element antennas have less directivity i.e. low gain in the particular direction as the radiation pattern of single element is relatively wide. Usually, high directivity i.e. more gain in the desired direction antennas required for long distance communications. High directivity antennas are possible by broadening the size of the radiating aperture which has maximum size larger than wavelength λ. But, this approach leads to the presence of multiple side lobes near the main lobe. And also, the antenna fabrication becomes difficult when the size of the antenna is large. There is an another method to increase the electrical size of an antenna is, an assembly of radiating elements in a suitable electrical and geometrical alignment is called an antenna array. Generally, all the elements in an array are identical. The design and fabrication of antenna arrays becomes simple by using identical elements in an array. The different discrete antenna elements are wire dipoles, loops, apertures, and etc. The vector combination of all the field patterns radiated by the individual antenna elements is termed as the radiation pattern. The more directive radiation pattern of an antenna array can be obtained by the fields produced by the individual antenna elements which are interfered constructively in one desired direction and then interfered destructively in the other directions [5].

A group of antenna elements whose current distributions are with various amplitudes and phases is referred as array antenna. To enhance the radiated signals in the particular direction and lessen it in the non-desired direction, the phenomena of electromagnetic wave is used. Antenna arrays are used to solve the limitations using

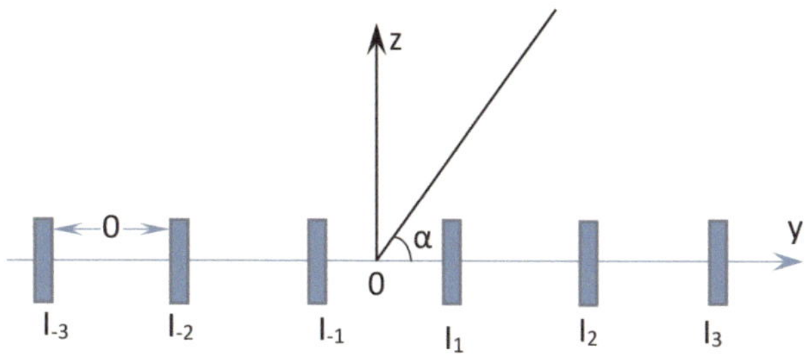

Figure 2.
Uniform linear array antenna.

single antenna. For example, a dipole antenna provides better radiation pattern compared to that of an isotropic antenna which radiates uniformly in all directions. But, the control of field pattern in a particular direction reduces when the length of the dipole increases. Therefore, controlling of radiation pattern by varying the lengths of single antenna is restricted. Hence, more flexibility and control over the gain can be attained for directing the antenna beam using multiple antennas/radiators arrangement (**Figure 2**).

1.1 Radiation pattern, antenna arrays and array factor

Antenna arrays of two types, i.e. one dimension and two dimension based on the arrangement of the antenna elements. Due to the simple construction of an array, one-dimension antenna array is used in this chapter. These arrays provide a specific radiation/field pattern. The total radiation pattern of the array varies when different antenna elements with different radiation pattern are combined [6]. This occurs due to array factor which measures the effect of joining radiating elements in an array without which the specific radiation pattern of each element is considered. Therefore, the complete radiation pattern of an array is calculated by this array factor and each antenna element radiation pattern. The whole radiation pattern results in a certain direction. Thus the efficiency and the directivity are calculated. These both are same if the overall efficiency is 100% [7].

Depending on the orientation of field pattern in a particular direction, arrays can be divided into broadside i.e. the radiation pattern is normal to the array axis and end fire i.e. the field pattern is parallel to the array axis. In this chapter, broadside arrays are taken into consideration and also radiation pattern in z direction is considered.

1.2 Defining array factor

Array factor is based on the number of elements in an antenna, spacing between the radiating elements, and also amplitude, phase of the signal applied to individual element. The overall surface area of radiating structure depends on the number of elements in an array and the spacing between elements. This area is termed as aperture. Large aperture area results high gain. The aperture efficiency indicates that how efficiently the aperture area is used. In this chapter, the influence of all these parameters are explained by considering linear array of isotropic radiating antenna elements [8]. An isotropic antenna is one which radiates power equally in all directions, i.e. isotropic antenna has unity directivity (0dB), unity gain (0dB) and efficiency of 100%.

The Array Factor is a function of the antenna element's position and the weights respectively [9]. By fitting these position and weight parameters, the performance of an array is optimized to get required properties. By changing the weights of elements, the antenna arrays can be steered in direction of maximum radiation pattern and the directivity can be increased with the increase in the number of antenna elements [10]. But due to the increase in number of elements increases the side lobes which are appeared next to the main lobe. From the definition of array factor there are two main lobes. One lobe is at $\theta = 0°$ and an another main lobe at theta = 180° in the directions of positive and negative x-axis respectively.

1.3 Thinned antenna arrays

One of the methods for optimizing antenna arrays around 1960 is known as array thinning. Thinning an array is simple and used in a uniformly spaced linear or planar array. Generally, building of large arrays with more number of elements are complex and also increased the fabrication cost. Therefore, removing some of radiating elements in an array would be desirable at which the performance is not significantly reduced.

The process of removing elements systematically from an array without significant degradation in the array performance i.e. the elements can then be disturbed from their exact positions if necessary. The main advantage of thinned array is the elimination of large side lobes easily. From the definition, the actual radiation pattern of an array is the multiplication of the Array Factor and the field pattern of individual antenna elements which make the array. So, by choosing an antenna with less gain and high angle, side lobes are removed instantaneously [11, 12]. Another advantage of thinning is that it does not need any mathematical computations. The array spacing uses non-uniform and sparse spacing. Thinning an antenna array comprises of the deduction (turning off) of radiating elements from an array [13] with reduced cost and weight of individual elements. This method of thinning permits closely the same narrow beam width which is equal to the filled array with total elements of equal size. In array thinning, lower side lobes can be acquired during turned ON process, the antenna elements operate with equal amplitude similar to the same filled array with uniform weighting.

One of the successful process of lowering the side lobe levels in an array is to reduce the magnitude of the weights of the corresponding elements from the center of the array [11]. This process is similar to the method of windowing in digital signal processing. Higher side lobes will appear when the array elements with uniform weights which are set across the array than the weights of individual elements are taper down. The density tapering requires uniform weights for all type of antennas. The elements in antenna array away from the center have less radiated energy compared to the remaining elements which are far away from the center. So, by removing these elements from the array will not decrease the performance of array.

As the computers are so computationally fast in these days, there is an another way of thinning and placement optimization has done using machine learning algorithms. All of these computerized methods employ statistical optimization techniques remove the antenna elements from an array without degrading the performance of an array [14]. The different machine learning algorithms like Support Vector Machine (SVM), Logistic Regression (LR) provide the enhanced thinned antenna array with high accuracy levels, reduced error and time while maintaining a feasible prediction of the antenna behavior, good computing efficiency with lesser number of iterations.

The steps to apply ML algorithms in the design of antenna arrays are explained here.

Step 1: Find the electromagnetic characteristics of an array Antenna using multiple simulations
Step 2: These simulated features are kept in the database and can be used as reference data for training using ML algorithms.
Step 3: Using these algorithms and making predictions array antenna provides the nearest results.

Support Vector Machines (SVMs) algorithm is used for the design of thinned arrays. This algorithm is used for regression, so it is termed as Support Vector Regressors (SVRs). This learning algorithm is trained with the dataset stored which includes computed values of the operational bandwidth, input impedance and the resonant frequency of the antenna array. It is shown that thinning of linear antenna array design using ML algorithms provides better and more accurate characteristics compared to theoretical. This section analyses the concept of thinning an array for lower side lobe sum patterns and also introduces these thinned arrays for difference patterns with low side levels. The Amplitude weights taken as either '1'or '0'. The antenna element is connected to matched feed input when the amplitude weight is '1'. When the amplitude weight is '0' the antenna element is connected matched load [11].

To get the improved performance of thinned arrays with the reduced side lobe levels, Mailloux and Cohen [12] proposed the 3-level amplitude weighting structure. This structure is used to approximate side lobe levels with Taylor amplitude distribution and statistical thinning process of the radiating elements. Hence, the thinning process in antenna array having been used more than 40 years and get good performance in military phased array radars. These phased array radars provide different operations like instantaneous wideband operation, mono pulse tracking and they are especially used for the recognition, tracking, and detection of intercontinental ballistic missiles [14]. Finally, thinned arrays in phased array radar systems provide high performance with low side lobe levels [15, 16].

2. Formulation

The amplitude is calculated for antenna arrays by considering different number of elements and these results are presented in **Tables 1** and **2.**

$$E(u) = 2\sum_{n=1}^{N} A(n)\cos\left[kd(n-0.5)(u-u_0)\right] \quad (1)$$

d = resonant spacing, ($\lambda/2$), k = $2\pi/\lambda$
A(n) = Amplitude distribution
The amplitude is also measured by considering different number of elements with Ishimaru spacings and these results are presented in **Tables 1** and **2**.

$$E(u) = \sum_{n=1}^{N} A(x_n)e^{j\left[\frac{2\pi L}{\lambda}ux_n+\phi(x_n)\right]} \quad (2)$$

Number of elements N	Amplitude distributions $A(X_n)$	Amplitude distributions $A(X_n)$ after Thinning for 2 &19 elements	λ/2 spacing	Ishimaru spacing X_n
1	1.00	1.00	0.5 λ	−0.95
2	1.00	0.00	0.5 λ	−0.85
3	1.00	1.00	0.5 λ	−0.75
4	1.00	1.00	0.5 λ	−0.65
5	1.00	1.00	0.5 λ	−0.55
6	1.00	1.00	0.5 λ	−0.45
7	1.00	1.00	0.5 λ	−0.35
8	1.00	1.00	0.5 λ	−0.25
9	1.00	1.00	0.5 λ	−0.15
10	1.00	1.00	0.5 λ	−0.05
11	1.00	1.00	0.5 λ	0.05
12	1.00	1.00	0.5 λ	0.15
13	1.00	1.00	0.5 λ	0.25
14	1.00	1.00	0.5 λ	0.35
15	1.00	1.00	0.5 λ	0.45
16	1.00	1.00	0.5 λ	0.55
17	1.00	1.00	0.5 λ	0.65
18	1.00	1.00	0.5 λ	0.75
19	1.00	0.00	0.5 λ	0.85
20	1.00	1.00	0.5 λ	0.95

Table 1.
Uniform amplitude distribution for N = 20 with different spacings.

Number of elements N	Gaussian amplitude distributions $A(X_n)$	Gaussian amplitude distributions $A(X_n)$ with Thinning of 2 & 19 elements	Ishimaru spacing X_n	λ/2 spacing
1	0.650	0.650	−0.95	0.5 λ
2	0.599	0	−0.85	0.5 λ
3	0.606	0.606	−0.75	0.5 λ
4	0.652	0.652	−0.65	0.5 λ
5	0.718	0.718	−0.55	0.5 λ
6	0.792	0.792	−0.45	0.5 λ
7	0.862	0.862	−0.35	0.5 λ
8	0.921	0.921	−0.25	0.5 λ
9	0.963	0.963	−0.15	0.5 λ
10	0.985	0.985	−0.05	0.5 λ
11	0.985	0.985	0.05	0.5 λ
12	0.963	0.963	0.15	0.5 λ

Number of elements N	Gaussian amplitude distributions $A(X_n)$	Gaussian amplitude distributions $A(X_n)$ with Thinning of 2 & 19 elements	Ishimaru spacing X_n	λ/2 spacing
13	0.921	0.921	0.25	0.5 λ
14	0.862	0.862	0.35	0.5 λ
15	0.792	0.792	0.45	0.5 λ
16	0.718	0.718	0.55	0.5 λ
17	0.652	0.652	0.65	0.5 λ
18	0.606	0.606	0.75	0.5 λ
19	0.599	0	0.85	0.5 λ
20	0.650	0.650	0.95	0.5 λ

Table 2.
Gaussian amplitude distribution for N = 20 with different spacings.

Using the above expressions, the radiation pattern of arrays are calculated
Here, $A(x_n)$ is the Amplitude distribution
$\frac{2L}{\lambda}$ = length of an array
u = sinθ, θ is the angle in the bore sight direction
x_n = Ishimaru spacing function

$$X_{n0} = (2n - N - 1)/N$$

Where $\phi(x_n)$ is the excitation of phase distribution.

Thinning an array comprises of the deduction (turning off) of radiating elements from an array [13] with reduced cost and weight of individual elements. This method of thinning permits closely the same narrow beam width which is equal to the filled array with total elements of equal size. In array thinning, lower side lobes can be acquired during turned ON process, the antenna elements operate with equal amplitude similar to the same filled array with uniform weighting.

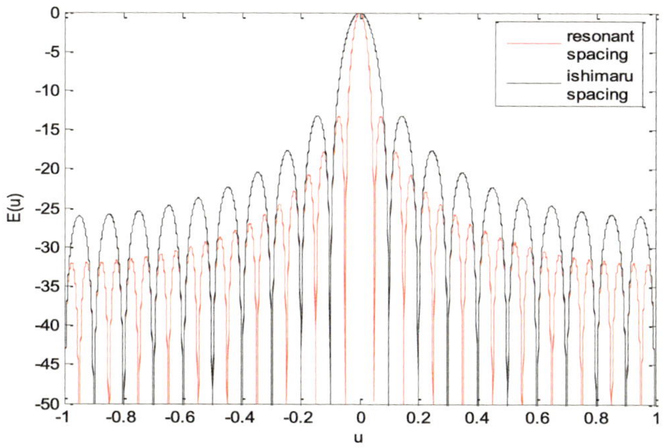

Figure 3.
Radiation pattern with resonant spacing and Ishimaru spacing for uniform amplitude distribution using N = 20 antenna elements.

Tables 1 and **2** gives Gaussian distribution and uniform amplitude distribution for N-20 elements using different spacing respectively.

Figure 3 presents the radiation pattern for N = 20 elements with resonant spacing and Ishimaru spacing for uniform amplitude distribution.

Figure 4 gives the radiation pattern for N = 20 elements with Resonant spacing and Ishimaru spacing for uniform amplitude distribution with thinning concept where the antenna elements i.e. '9' and '19' are eliminated from an array antenna.

Figure 5 gives radiation pattern for N = 20 elements with Resonant spacing and Ishimaru spacing for Gaussian amplitude distribution.

Figure 6 represents the Radiation pattern for N = 20 elements with Resonant spacing and Ishimaru spacing for Gaussian amplitude distribution with thinning concept where the antenna elements i.e. '9' and '19' are eliminated from an array antenna.

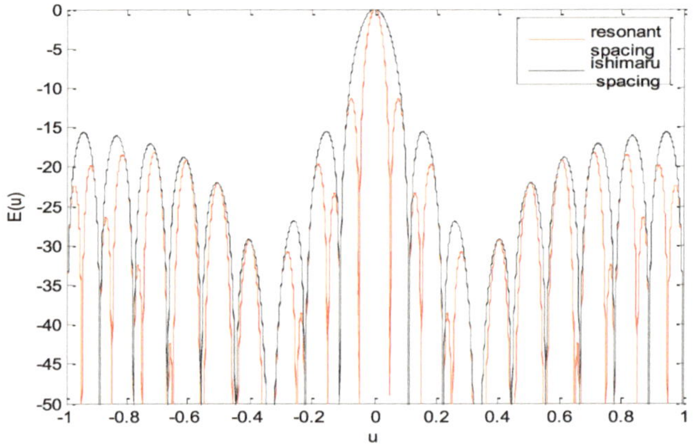

Figure 4.
Radiation pattern with Resonant spacing and Ishimaru spacing for uniform amplitude distribution with thinning using N = 20 antenna elements.

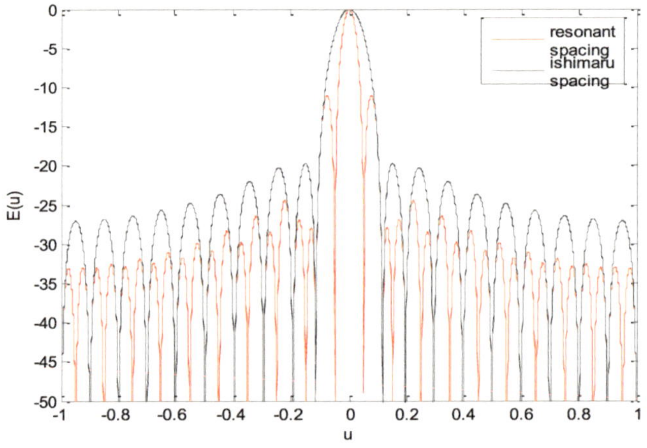

Figure 5.
Radiation pattern with Resonant spacing and Ishimaru spacing for Gaussian amplitude distribution. for N = 20 elements.

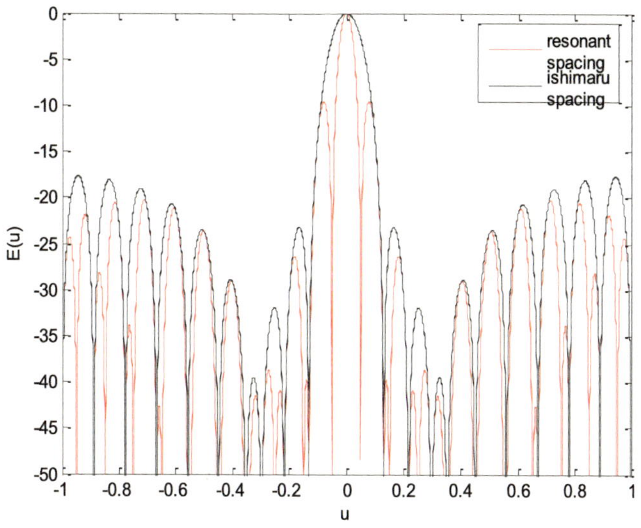

Figure 6.
Radiation pattern for N = 20 elements with Resonant spacing and Ishimaru spacing for Gaussian amplitude distribution with thinning.

3. Conclusions

The major advantage of present work is reduction of number of elements for getting specific patterns for the given array length by thinning. It is also more useful for obtaining the desired excitation levels at uniform levels while some of the elements are turned off. Here thinning concept is applied for resonant spacing and for Ishimaru spacing and observed the response for 2nd and 19th element turned off. Hence, the percentage of Thinning is 10% or percentage of Filling is 90%. From the results the beam width for Ishimaru spacing is more whereas the beam width for Resonant spacing is less and side lobe levels for Resonant spacing is more and the same response is observed for Gaussian distribution for array of 20 elements using machine learning algorithms. Such patterns are useful in radar applications to reduce EMI and creating nulls in unwanted directions.

Author details

M. Laxmi Prasanna Rani*, Moturi Satyanarayana and Narsupalli Shanmukha Rao
Department of ECE, MVGR College of Engineering, Vizianagaram, Andhra Pradesh, India

*Address all correspondence to: prassugowtham@gmail.com

References

[1] Razavi A, Forooraghi K. Thinned arrays using pattern search algorithms. In: Progress in Electromagnetics Research. PIER; 2008. pp. 61-71

[2] Russell SJ, Norvig P. Artificial Intelligence: A Modern Approach. 3rd ed. Upper Saddle River, New Jersey: Pearson Education, Inc.; 2016

[3] Mohri M, Rostamizadeh A, Talwalkar A. Foundations of Machine Learning. The MIT Press; 2012

[4] Harrington P. Machine Learning in Action. Manning Publications; 2012

[5] Skolnik M, Sherman J III, Ogg F Jr. Statistically designed density- tapered arrays. IEEE Transactions on Antennas and Propagation. 1964;**AP-12**(4): 408-417

[6] Lo Y, Lee S. A study of space-tapered arrays. IEEE Transactions on Antennas and Propagation. 1966;**AP-14** (1):22-30

[7] Raju GSN, Reddy RR, Satyanarayana M. Sum patterns from arrays of longitudinal slots. In: International Conference on Electromagnetic Compatibility. Bangalore, India: INCEMIC; 2010. pp. 335-339

[8] Linear arrays with arbitrarily distributed elements. IEEE Transactions on Antennas and Propagation. 1960; **AP-8**:222-223

[9] Unz H. Linear Arrays with Arbitrarily Distributed Elements. California, Berkeley: Electron. Res. Lab., Univ; 1956

[10] Chakradhar KS. Three different compact elliptical slot UWB antennas for wireless applications. Journal of Communications. 2021;**16**:52-59

[11] Keizer WPMN. Amplitude-only low Sidelobe synthesis for large thinned planar array antennas. IEEE Transactions on Antennas and Propagation. 2012;**60**:1157-1161

[12] Tomiyasu K. Combined equal and unequal element spacings for low sidelobe pattern of a symmetrical array with equal-amplitudeelements. IEEE Transactions on Antennas and Propagation. 1991;**39**(2):265-266

[13] Randy I, Haupt L. Fellow interleaved thinned linear arrays. IEEE Transactions on Antennas and Propagation. 2005;**53**:9

[14] Jiang R, Wang X, Cao S, Zhao J, Li X. Deep neural networks for channel estimation in underwater acoustic OFDM systems. IEEE Access. 2019;**7**: 23579-23594

[15] Rocca LP. Large array thinning by means of deterministic binary sequences. IEEE Antennas and Wireless Propagation Letters. 2011;**10**:9

[16] Randy L. Thinned arrays using genetic algorithms. IEEE Antennas and Wireless Propagation Letters. 1994;**42**:7

Chapter 4

The Issue of Sidelobe Level in Antenna Array: The Challenge and the Possible Solution

Onu Kingsley Eyiogwu

Abstract

This work looked at the issue of sidelobe level associated with antenna array and the challenge, providing a way that the issue and challenge can be overcome. Different authors have used different techniques in trying to find solutions to the problems in antenna array. This research work considered enhanced firefly algorithm and genetic algorithm as techniques or methods in addressing the issues in antenna array, such as side lobe level reduction. The work has shown that enhanced firefly algorithm (EFA) performs better than genetic algorithm (GA) in optimizing side lobe level of antenna array without any serious effect on the beam width. The best side lobe level obtained while using enhanced firefly was (−32.5 dB) compared to the value of (−20.0 dB) obtained with genetic algorithm. It was also found that tilting beam towards the direction of end fire, genetic algorithm optimization technique can be used to reasonably reduce the level of side lobe in array antennas. It was also found from this research work that reducing the number of active elements in an antenna array by way of turning OFF some elements can be used to improve performance of the array.

Keywords: antenna array, array factor, wireless communication, firefly algorithm, SLL

1. Introduction

The field of wireless communications has continued to expand. The authors in [1] noted that the expansion of wireless communication and GSM subscribers in recent times is astronomical. A key part of wireless communications technology is antenna. As the expansion of wireless communication technology continues globally, becoming a part of our day-to-day existence, problems brought about by such expansion require urgent and continuous attention. Experts and researchers have brought in a number of innovations, at least, to enable people freely enjoy the full benefit of ever-evolving wireless technology. Such innovations include the use of antenna arrays. By antenna arrays, I mean the connection of different antennas in such a way that the different antennas function as one lone antenna, enabling transmission only in the desired direction while suppressing it in any other direction.

The method of beamforming was also adopted in antenna arrays. Beamforming is a method in antenna arrays wherein a combination of antennas that forms an array is made to transmit signals towards a particular direction of interest instead of transmitting in all possible directions. In beamforming, the signal of each antenna that makes up the array will have its amplitude and phase adjusted so as to ensure that the transmitted signal is focused right onto the target beam. There are usually destructive and constructive effects associated with the combination of the different signals in beamforming arrays. Therefore, the radiation pattern does not have a single lobe but many of it, even different field strengths from different angles. So there exists the following: the main lobe which has the peak power, and it is the desired beam, the beam of interest; and the unwanted lobes that are smaller than the main lobes. The minor lobes do not radiate signals in the same direction as the main lobe but radiate in directions that are completely unnecessary. This is a very serious issue that requires attention so that the side lobe level could be reduced to the bearable minimum. The use of beamforming can be at both ends: at the receiving end and at the transmitting end. This is done primarily to obtain a spatial selectivity. When there is need for the detection and estimation of a particular signal of real interest, adaptive beamforming is used.

As mentioned, the arrangement of antennas to form arrays can be used for transmission or for reception. This means that instead of using a single antenna for transmission and a single antenna for reception, antenna arrays can be used for reception and transmission. Of course, the use of a single antenna for transmission or reception and an array antenna on the other side (either for reception or transmission) also exists. When multiple antennas are used at both the input (transmission) and output (reception), the technology is termed multiple input multiple out (MIMO) system. This arrangement is used to achieve better gain, more effective reliability in wireless communication [2], and for finding radio direction as pointed out in [3].

1.1 Literature review

The authors in [4] worked on optimizing planar and linear array antenna through the use of firefly algorithm. The authors described and reported three (3) case applications that show the effectiveness of firefly algorithm in achieving the anticipated optimization of array antenna, particularly planar array antenna. First, distribution of isoflux with linear array comprising of isotropic antennas that were not uniformly spaced had the radiation antenna synthesized. Then, radiators radiating equally in all directions and mounted on a nanosatellite that formed a planner array, though not uniformly spaced, was optimized. Lastly, the authors described an optimization process of a three by three planar array antenna used for the purpose of beam steering, having simultaneous level of side lobe. In [5] modeling of two annular ring array antennas, made up of elements radiating equally in all directions, was undertaken to generate isoflux radiation pattern to be used by satellites located in medium – earth-orbit or geostationary-orbit. To minimize the level of the sidelobe and shape the beam, the authors made use of differential evolution (DE).

In [6] the study of circular antenna array design and the design of concentric circular array antennas made up of isotropic radiators for the reduction of the level of sidelobe optimally was undertaken. Firefly algorithm was relied upon in finding the optimum position and set of weights for the circular antenna arrays and for the concentric circular array antennas which can give a pattern of radiation with the level of sidelobe really reduced. The authors noted that the firefly algorithm performed better than other optimization techniques like particle swamp optimization genetic algorithm (GA) etc.

A way of reducing poor convergence speed associated with firefly algorithm was investigated in [7]. The authors looked at minimization to a reasonable extent, of the sidelobe level without any serious consequence on the width of radiated beam. The results of the work were then compared to the result obtained using genetic algorithm. In [8] the authors carried out research to address the issue associated with synthesizing linear antenna array. Firefly algorithm was the optimization technique that the authors used in achieving their objective but with special emphasis on controlling the excitation of the amplitude of array element. A comparison of the firefly algorithm with Particle Swamp Optimization (PSO), Self Adaptive Differentials Evolution (SADE), and Tajuche Optimization method (TOM), with firefly algorithm showing better performance than others.

As can be seen in [9] a method which depended on evolutionary algorithm was used to synthesize rectangular antenna array pattern. Maximum excitation of antenna elements was noted by the authors. A total of thirty (30) isotropic antenna elements were considered, the spacing between successive elements was 0.5λ. Simulation results indicate that the peak sidelobe level was below 19 dB. In [10] a method of pattern synthesis that combined artificial bee colony and firefly algorithm was used to produce footprint patterns for a satellite based on rectangular planar array having antennas radiating equally in all directions. The authors carefully modified some key parameters like the phase, array elements state, and even the amplitude.
The performance of generic algorithm, firefly algorithm, and artificial bee in getting the optimum solution to the wanted footprint patterns was also compared, with artificial bee colony and firefly algorithm seen to perform better than genetic algorithm.

Mandal et al. [11] presented a method or technique that uses differential algorithm (though of a single objective) to minimize a multi-objective fitness function. In the work, conflicting parameters were optimized. Such conflicting parameters were low maximum level of sideband radiation, low value associated with maximum level of sidelobe, and the main beams narrow beamwidth. The proposed method was then applied to a uniformly excited time-modulated linear antenna arrays and non-uniformly excited time-modulated linear antenna arrays in the synthesis of optimum pattern of low sidelobe at the frequency of operation. This was achieved by the suppression of the level of radiation in the side.

1.2 Description of antenna array

Antenna array is the arrangement of different antennas into a single one, designed to be used for a particular purpose. Antenna array comes in different shapes and sizes; hence, we have the linear antenna array, the rectangular, the planar and so forth. The size of antenna arrays is basically determined by the number of individual elements or antennas combined to form the array. Antenna array shape, distance between successive antenna elements, the amplitudes of excitation, and of course, element phase excitation are some of the factors on which array antenna factor depends. To have an optimized antenna array, [7] noted that the distance between the elements of the array has to be controlled. Controlling radiation pattern of antenna array is achieved by the amplitudes and phases of current excitation. Reduction of sidelobe and suppressing interference can all be realized by a careful control of the key parameters as mentioned above.

Some experts classify antenna arrays into two, basing the classification on how the axis of the antenna relate to the radiation direction. Therefore, there are endfire array and broadside array. Endfire array is usually linear, having its radiation direction being the same as the line of the antennas. Normally, in endfire array, the phase difference by which the antennas are fed is equal to the distance between two adjacent antennas. However, the feeding of the antennas in broadside array is done in phase. This is to ensure that there is a perpendicular radiation of radio waves.

Of course, other types of antenna arrays exist, and they include, but not limited to:

- Parasitic arrays
- Turnstile array
- Yagi-Uda array
- Parasitic array
- Collinear array
- Rectangular array
- Phased array

Phased array is the category of antenna array that is designed with the capability of changing the shape and direction of radiated signal through electronic steering without having to move the antenna physically. The electronic steering is made possible by the difference (in phase) existing between the various signals coming from the antennas making up the array. Signals radiated by the antennas in phased array can either be in phase or out of phase. When they are in phase, the signals are added, resulting in additive signal amplitude. When the signals are however out of phase, the signals are seen to cancel out each other.

Phased array antenna has three basic types, namely, linear array which has the elements positioned on a straight line with only one phase shifter; Frequency scanning antenna array which does not have any phase shifter at all but the transmitters' frequency is used for beam steering; and the planar array which has the elements arranged in planar form, with each antenna having a phase shifter.

Despite great advantages and benefits of antenna array, there are obvious issues and challenges that need to be addressed to enable people enjoy the full benefit of wireless communications technology. In rectangular array, the issue lies with the level of side lobe which has to be reduced. There are also issues in array synthesis such as computational cost, especially as the size of the antenna increases. Experts have continued to search for possible solutions to the lingering problems associated with antenna arrays, using different methods and techniques.

The use of rectangular antenna array provides far better advantage than using single individual antenna elements in wireless communications. The advantage of using rectangular antenna array over the use of single antenna element in wireless communications include the realization of low sidelobes, higher directivity of symmetrical patterns [12].

A rectangular antenna array can have M number of antenna elements in one direction and N number of antenna elements in the adjacent direction, giving rise to M by N rectangular array antenna elements (M x N).

The array factor computation for such M by N rectangular array takes into consideration the directions of the M elements and N elements, and it is given by [12].

$$AF = AF_M AF_N \tag{1}$$

Where AF_M is the array factor in the M direction and AF_N is the array factor in the N direction.

$$AF_M = \sum_{m=1}^{M} I_{m1} e^{j(m-1)(kd_M Sin\theta Cos\varnothing + \alpha_M)} \tag{2}$$

$$AF_N = \sum_{n=1}^{N} I_{N1} e^{j(N-1)(kd_N Sin\theta Cos\varnothing + \alpha_N)} \tag{3}$$

$$\text{K represents phase constant, and it is given by } k = \frac{2\pi}{\lambda} \tag{4}$$

α_M = elements phase shift in the M direction.
α_N = element phase shift in the N direction.
θ = zenith angle.
$\varnothing$ = azimutal angle.
α_N = element progressive phase shift in N direction.
α_M = element progressive phase shift in M direction.

If the rectangular array is symmetrical the computation of array factor is given by [3].

$$\mathrm{AF} = 4\left[\left[\sum_{m=1}^{M}\sum_{n=1}^{N/2} Cos\ (m-0.5)(kd_m U + \alpha_m)\right)\ Cos\ (n - 0.5\ (kd_N(kd_m V + \alpha_N))\ \alpha_m)\right) \tag{5}$$

In Eq. (6) above, v = sin $\varnothing$ sin θ while U = Cos $\varnothing$ Sinθ.

1.3 Issues in antenna array

Issues in antenna array include, but not limited to, the following:

- High level of side lobe
- Array synthesis problems
- Thinning issues
- Cost, power consumption etc.

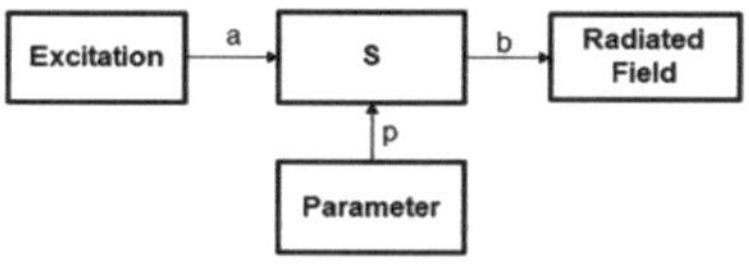

Figure 1.
An antenna system model.

One major issue in antenna array has to do with the synthesis of the array. When the specifications for antenna array design are provided, such as the structure of the required array, the number of required radiating elements, radiated pattern required and so forth, determining the exact excitation and the feeding network that gives such excitation and determining structure of the array that satisfy the requirements of the design, is a key problem. In [7] it was stated that algorithm for synthesizing array is more or less a minimization one.

Without considering beamforming network, antenna system can be conveniently represented as shown in **Figure 1** above. From **Figure 1**, p and a are the inputs which represent the required parameters with their electromagnetic and geometric properties respectively, b denotes output, which is the radiated field [7]. S denotes the operator, which is dependent on the frequency. It should be noted that while S is linear in a, it is non-linear in p. Based on **Figure 1**, b, p, and a denote the sharp constraint in the radiated field, in the parameter as well as excitation respectively; then the minimum of Eq. (6) needs to be found if the problem of antenna array has to be solved.

$$d^2 = inf\ //\mathrm{y}–y_c//2 \tag{6}$$

$$\mathrm{Xn} = \beta_0 e^{-rd^2 i,i}\left(x_j–x_i\right) + \mathrm{L(rand)} \tag{7}$$

Note: L(rand) refers to the randomization based on levy flight, x_i and x_j represents the positions of i and j fireflies respectively.

$$L = 0.01\left(\frac{d, \propto}{/d_2//^{\frac{1}{\beta}}}\right) \tag{8}$$

Where

$$\propto = \left(\frac{\Gamma(1+\beta)\, sin\left(\frac{\beta\pi}{2}\right)}{\Gamma\left(\frac{\beta+1}{2}\right)\beta^2\left(\frac{\beta-1}{2}\right)}\right)^{\frac{1}{\beta}} \tag{9}$$

The coefficient of absorption, Γ, is used to determine speed of convergence.

1.4 The array

To fully understand the challenge involved in antenna arrays, a quick look at Eq. (6) will prove helpful. Addressing problems of synthesis based on Eq. (6) has to do with finding a certain point where the set y is closest to $\mathrm{y_c}$. However, practically speaking, y and $\mathrm{y_c}$ are non-convex, showing that there will be points of relative minimum and absolute minimum, and minimization algorithm is entangled in a local minimum [13], thereby providing a wrong solution. The enlargement or trapping of the minimization algorithm is of key interest. Because as the size of antennas increase, number of secondary minima rises, and distinguishing between relative minimum and absolute minimum becomes rather difficult. Experts have therefore adopted algorithms for global minimization in order to solve the problem. In [13] it was emphasized that global optimization algorithms, practically stressing, do not provide any guarantee of realizing the optimum, especially with large problem size.

In any antenna array for improved performance, the challenge lies in getting an optimal set of spacing for the elements that satisfies the specification for the array, which of course must be based on how current is distributed with the elements of the antenna. When antenna array requires special radiation or when an off normal scanning is necessary, optimal thinning of antenna array gets highly difficult. Not to be forgotten is the fact that having array patterns optimized with respect to the location of the antenna element is non-linear as well as complex because the thinned array's array factor is not a linear function as far as element spacing is concerned.

There are also challenges like:

- Expansion of solution space
- Complexity of landscape of solution space

1.4.1 Expansion of solution space

The moment the size of the antenna is increased by increasing the number of antenna elements, solution space definitely expands to a large degree [14]. This implies, even from common sense, that searching exhaustively for solution is rather far more practical if the arrays are small.

1.4.2 Complexity of landscape of solution space

As mentioned, solution space increases as the antenna elements increase. And when this happens, there is every likelihood that the space for the landscape of solution will become more complex.

1.5 Thinning of antenna array

In array antenna, all the elements of the array are active. This makes the array performance to be degraded to some extent. However, some elements of the array can be made passive or be turned OFF. Hence, we have arrays that are fully populated, having all the elements active, and arrays that are partially populated, having some elements terminated to a well-matched load. The turning OFF of some elements of the array is termed thinning. Therefore, in thinned array, some elements are ON while some are OFF; in fully populated array all the elements are ON.

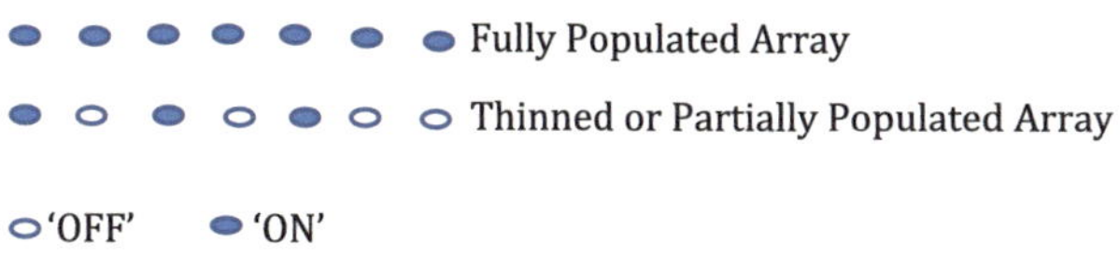

1.5.1 Genetic algorithm-based thinned phase Array

One major problem in phased array antenna is the level of side lobe which has to be reduced. Genetic algorithm is capable of optimizing the least level of the sidelobe. This is accomplished through the optimization of the OFF position and the ON position as regards the array antenna elements. As given in [15], the array factor for array antenna is:

$$AF = \sum_{1}^{n} I_n . e^{jn2\pi r(cos\theta - \cos \varnothing)} \tag{10}$$

Where n = number of elements and r = distance between elements. When using genetic algorithm for the optimization, Eq. (10) forms the cost function.

2. Solution to antenna array problems

Side lobe level, which is one of the key parameters to be minimized for effective performance of the antenna arrays, can actually be optimized or reduced in such a way that the system performance will not be adversely affected. Here, the use of enhanced firefly algorithm is considered.

2.1 Side lobe level (SLL) optimization using enhanced firefly algorithm

Though the use of enhanced firefly algorithm for effective reduction of side lobe level in antenna array will be presented here, it is however necessary to briefly describe firefly algorithm and state the reason for the choice of enhanced firefly algorithm instead of firefly algorithm itself.

2.1.1 Firefly algorithm

Humans have learnt so much from the natural world and they have applied such lessons in addressing some problems that face us. The behavior of some living things have been studied extensively and key aspect of their life mimicked by humans. One of such living things is the firefly, an insect that emits light, especially at twilight, to get the attention of other fireflies for matting. It uses that method to prey on other fireflies.

Experts mimic the way firefly flashes light to attract other fireflies to solve problems that are not linear, among others. Hence, there is firefly algorithm (FA) which rely on certain assumptions like: attraction between fireflies is independent of their sex: brightness of firefly is determined by objective function (OF) for such firefly algorithm, the most important things being the light intensity denoted by I and the attractiveness denoted by β. The extent of a firefly's brightness will determine how attractive it will be. The following equation gives the light intensity.

$$\mathrm{I} = l_o e^{-yd^2} \tag{11}$$

Where l_o is the initial light intensity, y is coefficient of absorption of light while d is the distance between two given fireflies. Similarly, the attraction is given by:

$$\beta = \beta_o e^{-yd^2} \tag{12}$$

Where β_0 is the attractiveness when d = 0. One major drawback of firefly algorithm is that it takes longer to attain global optimization if the array is bulky. To address this drawback, the firefly algorithm is improved, giving rise to improved or enhanced firefly algorithm (EFA).

2.1.2 Enhanced firefly algorithm

Enhanced firefly algorithm is adopted to address the problem of slow convergence, thereby overcoming the challenge of optimizing sidelobe level without serious consequence on the beam width.

A careful implementation of the flowchart for firefly algorithm can bring about real result in terms of reduction of sidelobe levels. Equations (11 and 12) are very important in the application of enhanced firefly algorithm (**Figure 2**).

2.2 Solution to thinning problem in antenna array

Addressing thinning issues also solves the problem of cost and power consumption, as these issues come about because of the bulky nature of some antenna arrays. One method that can be used to address the thinning problem associated with antenna array is the genetic algorithm. Compared to some frequently used techniques or methods, genetic algorithm offers great advantages owing to its robustness nature. Key advantages of genetic algorithm include the following:

- It performs well even with large variable numbers
- It does not need information of any kind on derivative
- The search from a spacious sampling involving the cost surface is done simultaneously
- It is very capable of working with experimental or generated data
- It does not provide a single solution but a group of optimum parameters
- Its optimization can be done with discreet or continuous parameters

One version of genetic algorithm that can be applied to solving antenna array thinning problems is the simple genetic algorithm (SGA) version. As the name implies, application of this version of genetic algorithm is actually simple. The steps, in the proper order, are as listed below:

- Coding of the parameters right onto the genes. This coding simply means a move from one space to another space such as from the parameter space to the one of the chromosomes. This coding also provides the transformation of the coded parameters to a string length of coded genes.
- Creation of initial population. This initial population is taken randomly, and it forms the chromosomes' starting matrix with its matrix elements which can either be in floating point, binary or floating point and mixed binary.

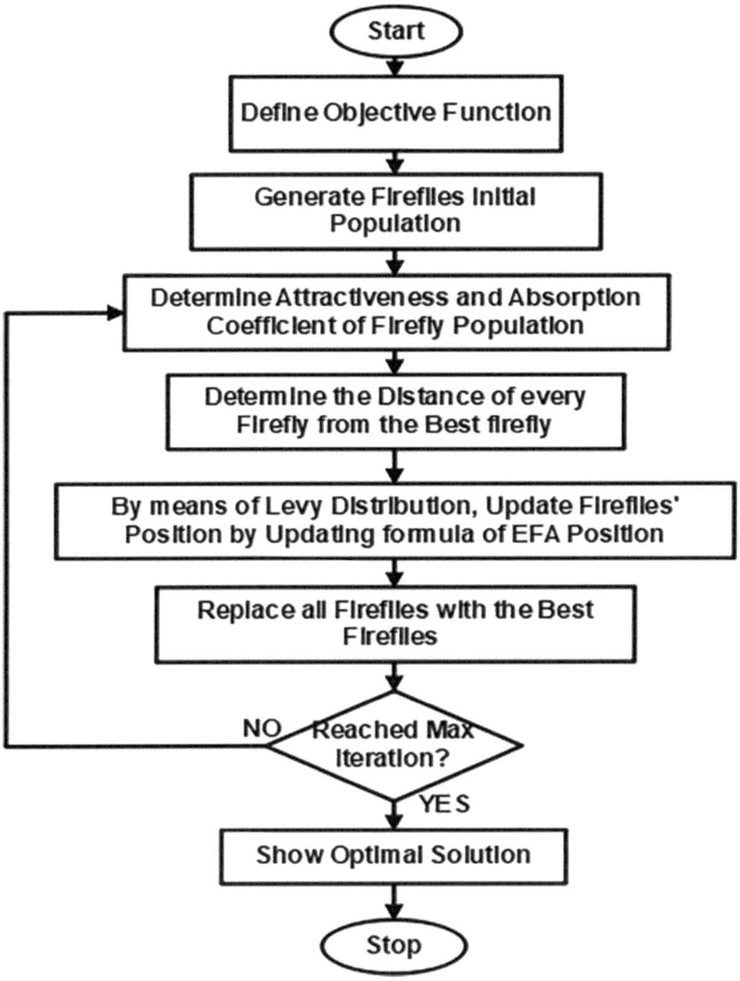

Figure 2.
Flowchart of enhanced firefly algorithm.

- Evaluation of fitness. Here, the evaluation of the original matrix to ascertain its suitability is carried out. Depending on the problem, the choice of the cost function is then made. It should be mentioned at this point that where the optimization is multi-objective, then normalization of each cost and separately weighing before a combination is undertaken to give one scalar quantity.

- Natural selection. Under this stage, the theory of survival of the fittest is applied. Two key methods exist for using natural selection: reject every other chromosome while retaining healthy ones; all chromosomes above certain value already predetermined are retained while others are discarded.

- Selection of a mate. Mating is carried out among the best members of the actual population. This mating is usually on probability. Tournament selection approach and that of Roullete Wheel are some of the very key approaches that are applied under selection of a mate.

- Off-spring selection. Crossover and mutation are main operators upon which the off-spring are produced. Mutation has to do with inducing variations in the population while crossover uses the parents to produce off-spring.

- Criteria for termination. The process of termination repeats until a criterion for termination is met. This happens when either the required iterations or the cost is satisfied.

2.3 Presentation of results and discusion

This part of the research deals with the results of the work carried out, and the discussion is based on the result. Figures are presented and discussed accordingly.

As can be verified from **Figure 3**, side lobe level is smaller for the optimized thinned array when compared with the level of the sidelobe for a fully populated array. This shows that when optimizing sidelobe level for antenna array using the method of thinning, genetic algorithm can be used for such optimization. The plot shows that for a fully populated array at an angle of 90°, the sidelobe level was −12.967 dB while for the thinned array optimized with genetic algorithm for the same angle of 90°, the level of sidelobe reduced to −16.857 dB.

As **Figure 4** shows, in thinned array optimized with genetic algorithm, the half-power beamwidth is higher compared to the level for fully populated array. While the half power beamwidth for thinned array optimized with genetic algorithm was 60° from **Figure 4**, for fully populated array the half power beamwidth was about 54.5°. This is an indication that when all the elements of an antenna array are ON, the gain is not the same as when some elements of the array are OFF. Thinning, therefore, is a key way of overcoming some challenges in antenna array, especially if thinning is optimized with genetic algorithm.

In **Figure 5**, the radiation pattern for antenna array of ten elements separated 0.5λ apart, having sidelobe level optimized using genetic algorithm is compared with the radiation pattern for antenna array of ten elements separated 0.5λ apart, having sidelobe level optimized using enhanced firefly algorithm. The plot shows that using enhanced firefly for sidelobe level optimization in antenna array is better than using the popular genetic algorithm. This is because optimizing using enhanced firefly has lower sidelobe level compared to optimizing using genetic algorithm.

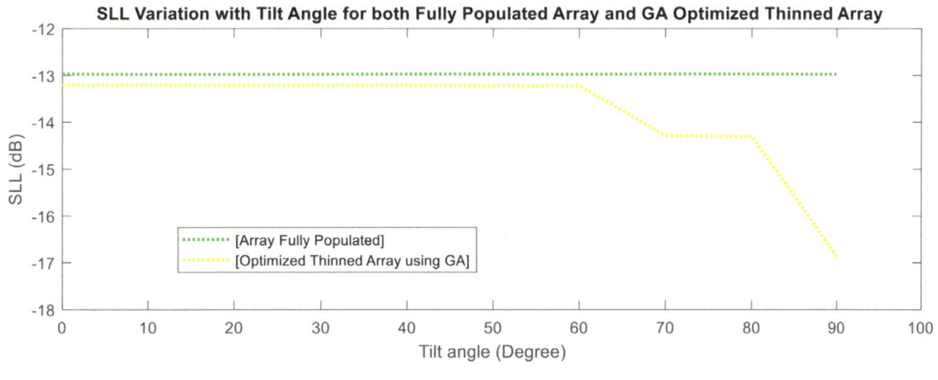

Figure 3.
A plot of SLL(dB) variation against tilt angle (deg.) for fully populated Array and optimized thinned Array using genetic algorithm.

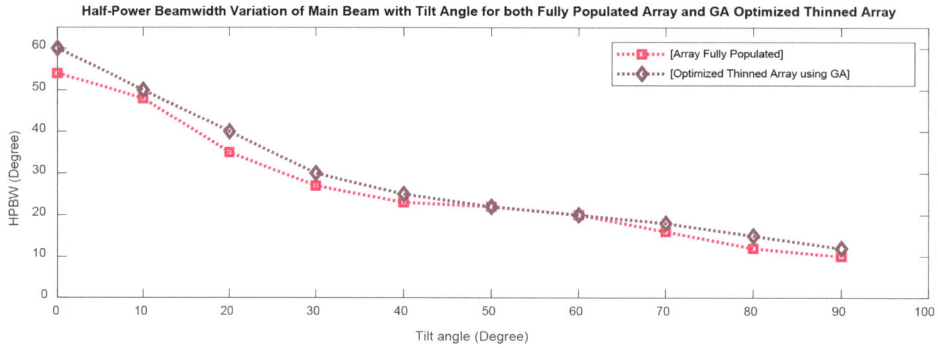

Figure 4.
A plot of half-power beam width (deg.) variation against tilt angle (deg.) for fully populated Array and optimized thinned Array using genetic algorithm.

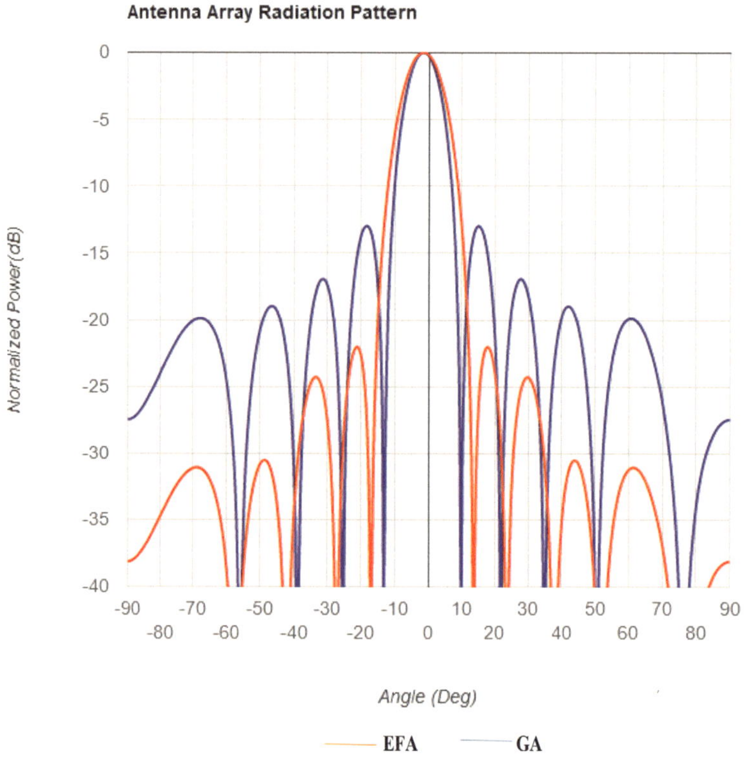

Figure 5.
An Array radiation pattern of ten elements spaced 0.5λ optimized with EFA.

Presented in **Figure 6** is the radiation pattern for antenna array of ten elements separated 0.5λ apart, with normalized power plotted against radiation angle in degree. On one hand, the sidelobe level is optimized using genetic algorithm and on the other hand, the sidelobe level is optimized using enhanced firefly. The result of the optimizations using genetic algorithm and enhanced firefly algorithm are compared. Again, the plot reveals that using enhanced firefly for sidelobe level optimization in antenna array is better than using genetic algorithm. The reason for this inference is because

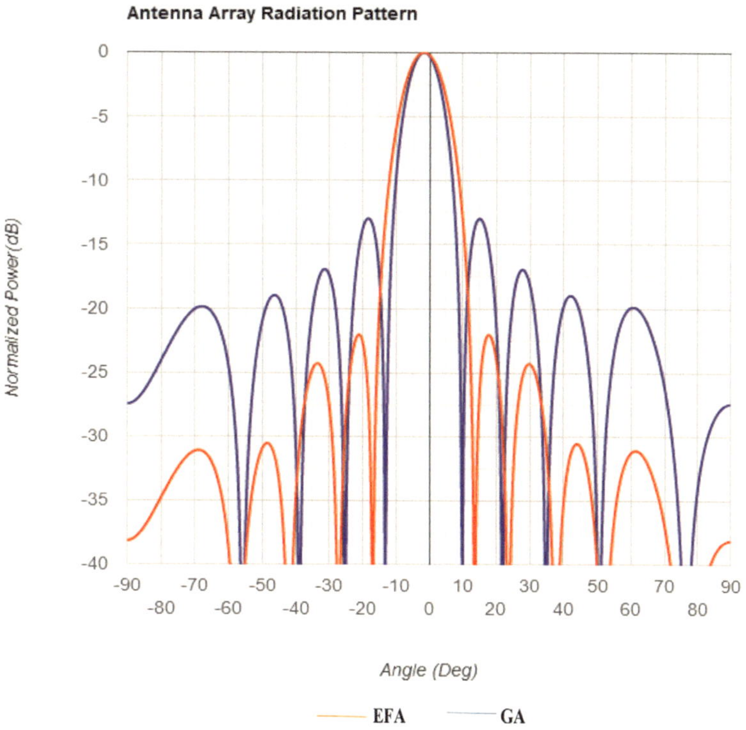

Figure 6.
A plot of normalized power (dB) against angle (deg) for 10 elements optimized with EFA.

optimization of sidelobe level using enhanced firefly has lower side lobe level compared to optimization using genetic algorithm.

3. Conclusion

This work looked at the issue of sidelobe level in antenna array and the challenge in reducing the sidelobe level, providing a way the issue and challenge can be overcome. Related literatures were reviewed, different antenna arrays listed, issues and challenges presented, and solution to the issue of sidelobe level outlined, with graphs. This work has shown that enhanced firefly algorithm (EFA) performs better than genetic algorithm (GA) in optimizing sidelobe level of antenna array without any serious effect on the beamwidth. Since sidelobe level can be optimized through phase or amplitude excitations, it is found that randomly-amplitude excited elements of rectangular antenna array have good sidelobe level than randomly-phase excited elements of rectangular antenna array. The best sidelobe level obtained while using enhanced firefly was (−32.5 dB) compared to the value of (−20.0 dB) obtained with genetic algorithm. As a way of improving this or similar work, the author suggests that a comparison of enhanced firefly algorithm with all other key optimization methods or techniques should be undertaken.

It was also found that tilting beam towards the direction of end fire, genetic algorithm optimization technique can be used to reasonably reduce the level of sidelobe in array antennas. Equally worthy of note, as found from this work, is the fact that reducing the number of active elements in an antenna array by way of turning OFF some elements can be used to improve performance of the array.

Nomenclature

AF	Array factor
EFA	Enhanced firefly
FA	Firefly algorithm
GA	Genetic algorithm
MIMO	Multiple input multiple output
PSO	Particle swamp optimization
SADE	Self-adaptive differential evolution
SGA	Simple genetic algorithm
SLL	Side lobe level
TOM	Tajuche optimization method
α_M	elements phase shift in the M direction
α_N	element phase shift in the N direction
θ	zenith angle
∅	azimuthal angle
α_N	element progressive phase shift in N direction
α_M	element progressive phase shift in M direction

Acknowledgements

I wish to thank my wife and children, Onu Faith, Onu Kingdom, and Onu Treasure for the encouragement I received from them to continue this important work. The Almighty God is highly appreciated for the wisdom to carry out this work.

Author details

Onu Kingsley Eyiogwu
Rivers State University, Port Harcourt, Rivers State, Nigeria

*Address all correspondence to: Kingsleyonu92@yahoo.com

References

[1] Elechi P, Orike S, Onu KE. Cellular Planning of GSM Network in Rivers State, Nigeria. Journal of Telecommunication, Electronic and Computer Engineering. 2020;**12**:44-53

[2] Poole I. What is MIMO? Multiple input Multiple Output. Antennas and propagation. Available from: radio-elec tronics.com [Accessed: April 18, 2022]

[3] Bevelacqua P. Array Antenna. Available from: https://www.antennathe ory.com [Accessed: April 18, 2022]

[4] Eduardo Y, Marcos VTH. Optimization of Planar Antenna Arrays using the Firefly Algorithm. Journal of Microwaves, Optoelectronics and Electromagnatic Application. 2019;**18**: 35-37. DOI: 1590/2179-107420gr18:11646

[5] Ibarra M, Andrade AG, Panduro MA, Mendez A. Design of Antenna Arrays for Isoflux Radiation in satellite Systems. In: Proceedings of 33rd International Performance, Computing and Communications Conference. Austin, Tx: IEEE; December 2014

[6] Ashraf S, Nihad D. Circular Antenna Array Synthesis using Firefly Algorithm. Department of Electrical Engineering, Jordan University of Science and Technology. UK: Wiley; 2013. DOI: 10.1002/mmce.20721

[7] Bhagya KNKS, Venkateswara NR. Optimization of SLL of Linear Array Antennas using Enhanced Firefly Algorithm. International Journal of Engineering research and Application. 2020;**10**:19-23

[8] Kamaldeep K, Vijay KB. Optimization of Linear Antenna Array using Firefly Algorithm. International Journal of Engineering Research & Technology (IJERT). 2013;**21**:2307-2313

[9] Debasis M, Ved PR, Anirban C, Anup KB. Synthesis of Dual Radiation Pattern of Rectangular Planar Array Antenna using Evolutionary Algorithm. ICTACT Journal on Communication Technology. 2018;**6**:1146-1149

[10] Chatterjee A, Mahanti GK, Ghatak G. Synthesis of Satellite Footprint Patterns from Rectangular Planar Array Antenna by using Swarm-based Optimization Algorithms. International Journal of Satellite Communication and Networking. 2013; **32**:25-47. DOI: 10.1002/sat.1055

[11] Mandal SK, Mahanti GK, Ghatak R. Differential Evolution Algorithm for Optimizing the Conflicting Parameters in Time-Modulated Linear Array Antennas. Progress in Electromagnetic Research B. 2013;**51**:101-118

[12] Venkateswara RN, Chenchu RR. Analysis of Rectangular Planar Array with Different Distributions and Optimization of Selected Level using GA and PSO. International Journal of Engineering Research and Technology (IJERT). 2015;**4**:577-581

[13] Jain R, Mani GS. Solving Antenna Array Thinning Problem using Genetic Algorithm. Applied Computational Intelligence and Soft Computing. 2012; **2012**:1-14. DOI: 10:1155/2012/946388

[14] Bucci OM, Urso MD and Isernia T. Some Facts and Challenges in Array Antenna Synthesis Problems. Available from: https://www.some-facts-and-cha llenges-in-array-antenna-synthesis-prob lems [Accessed: April 20, 2022]

[15] Banalis CA. Antenna Theory and Design. 3rd ed. Hoboken, NJ: John Wiley; 2015

Chapter 5

Coverage Determination of Incumbent System and Available TV White Space Channels for Secondary Use in Ethiopia

Habib Mohammed, Tessema T. Terefe and Sultan Feisso

Abstract

Different path loss modelsare used to analyze the behavior of terrestrial television signals. The path loss calculated by one model differs from the other depending on different factors they consider. Frequency is one of the main factors included in each model. The frequency variation in the electromagnetic spectrum causes different response for each model. In terrestrial TV signal representation, since it is operating under VHF and UHF spectrum range, the propagation model used to model the signal must be less invariant when the transmitter is operating in VHF and UHF. If the pathloss model used is very variant it is difficult to define the coverage of the transmitters. This causes interference among transmitters and between the digital terrestrial TV transmitters and TV white space devices. Different propagation models are analyzed by their sensitivity to frequency variation from very-high and ultra-high frequency spectrums. After the best model is selected, we have used this model to find the coverage of the incumbent transmitter, which then is used to analyze free channels for secondary use. First the pathloss at VHF and then for UHF is calculated. This difference is then compared and the result indicates that ITU-R P.1546–5, which incorporate terrain data is best of others. Using this model and further analyze the coverage and free channels, we have found a minimum of 408 MHz free contiguous bandwidth, by considering a worst-case scenario, which is placing a WSD at the incumbent transmitter.

Keywords: Contour Coverage, Free Channels, Pathloss Models, Co-Channel, Adjacent Channel, Field Strength, Digital Terrestrial TV, TV White Space

1. Introduction

TV White Space is the free channel or frequency range which was already dedicated for terrestrial television transmission in one country's boundary. The channels reserved for this primary use may be found free in time and spatial variations. The availability of free channels depends on these variations. The variations on the other hand resulted from geographical difference and other parameters. These parameters are factors on which different propagation models depend. Considering different parameters, the signal of TV propagation must be represented more closely to the natural signal loss of the propagation. For this comparison, different propagation models are taken in to account which have different response for different parameters. One may be very sensitive for frequency, where the other considers mainly the terrain and the other may be more sensitive to weather. In addition to these differences in consideration of different parameters, there are some parameters to affect all the models. Frequency is one of these parameters. The propagation losses have different values of sensitivity. Different sensitivity for frequency will cause different response for differences in frequency from very-high frequency (VHF) to ultra-high frequency (UHF). The difference is also not the same for the same variation in frequency for the two frequency ranges [1]. The availability of free channels directly related to the coverage of primary digital terrestrial TV (DTT) transmission system [2–5]. The coverage of incumbent system is directly related to pathloss models which are used to represent the signal propagation of TV signals. Its area of coverage is calculated by using the minimum receivable signal by the DTT receiver which is to be modeled with precise pathloss model [6]. If the pathloss model used for this modeling is very sensitive for frequency in VHF and UHF, it shows a significant variation in coverage area which may result in difference in number of available channels. So, we must select less sensitive model. Propagation models used for modeling the DTT signal, and also for tv white space (TVWS), must be perfect in operating under VHF and UHF band. Also, they must be models which can cover long range in signal propagation, since TV signals cover long distances with small power. In this paper, related works are revised in part II. The next part focuses on coverage determination factors. In IV, system model and problem formulation are discussed. After this, results are discussed and relevant conclusion is given then.

2. Related works

Many papers are done on propagation models to represent TV signals. These papers tried to compare different models with different parameters [7–11]. In [8] Irregular Terrain Model (ITM) is compared with Irregular Terrain With Obstruction Model (ITWOM) and it suggests ITM model to be better for TV signal modeling, since ITWOM is better in shorter distances which range up to 20 km and since TV signal covers distances up to 100 km and more. This paper also compares ITM, ITU-R P.1546–5 recommendation, Hata Devidson Model, Deygout, Episton-Peterson, and Giovanelli models with real measured data and points out that ITU-R P.1546–5 recommendation shows significant errors for distances above 50 km where as Hata Devidson model gives small errors. It suggests ITM to be better in error minimization with cost of huge computation time and steps, by considering terrain of some part of Greece. R. Gorrepati et.al in [9] analyzed the performance propagation models in estimating the TV coverage and they compared Hata and ITU-R P.1546–4 recommendation. The basic parameter for comparison is consideration of terrain data and the result shows the coverage is better when terrain data is considered using ITU-R P.1546–4 model. The paper concludes, propagation models which consider terrain data have better performance in coverage estimation TV transmission. But it only considers two models. Mesele Mekonen [7], tried to select propagation model for signal representation of TV signals. He has selected ITU, Stanford University Interim, Cost231 Hata, Okumura Hata, Okumura and Free space pathloss models for comparison. His comparison was by pathloss values of different propagation models and selected Okumura Hata pathloss model to have the lowest value for rural and for urban areas, SUI model is selected for shorter distances and Okumura model for longer distances. But according to [12], COST231 Hata model is valid for frequencies 500 MHz–1500 MHz and link distances up to 20 km. So, it does not include VHF frequency and long ranges of coverage. Also, Stanford University Interim according to [13], and ITU-R P.1411 model according to [14] are not used to model TV signals in both VHF and UHF range and for transmitter antenna heights of higher value. This paper does not also consider the real terrain map of Ethiopia. The propagation models selected for comparison must be valid for both VHF and UHF frequency ranges, larger range of signal coverage and higher Tx antenna heights. Papers done before do not try to compare the propagation models according to their sensitivity to frequency variation in VHF and UHF bands. Since this will cause a significant coverage difference in incumbent system, we have tried to compare pathloss models using their sensitivity behavior to frequency by including propagation models which include Ethiopian terrain data. And using this model, we have investigated the how much capacity is available in Ethiopia.

3. Coverage determination factors

3.1 Incumbent coverage determination

The coverage of primary DTT signal is the area between the transmitter and the points at which the received signal has the minimum receivable quality. At these points the signal power is considered to be of minimum value for a primary receiver below which it cannot have a viewable quality [2]. The determination of coverage area is essential in primary system deployment in order to avoid interference [3]. It can be determined by using noise limited contour or interference limited contour. The former method formulates the coverage area to be the area under the points where the signal carrier to noise ratio is less than the difference of minimum receivable signal value and noise floor. The later method is based on similar formulation where it uses interference instead of noise floor [15].

3.2 Terrestrial TV network frequency

The cellular system for DTT planning uses different planning system [2]. The frequency assignment for each transmitter can be of single channel or multiple channels. If the transmitter broadcasts with two or more channels in its coverage range, it is multiple frequency transmission. Otherwise, it is single frequency transmission. For Ethiopian case, the responsible body, Ethiopian Broadcasting Authority, plans which form of transmission should be deployed in one place. Dominantly, multiple frequency transmission is deployed in the country. The authority is also responsible for other planning strategies for terrestrial TV. The single frequency transmission uses only one frequency and there must be sharp gap between neighboring cells in order to avoid interference. It is usual to use in digital TV transmission. Multiple frequency transmission on the other hand relies on using different channels in the licensed range of frequency for TV all over the country. Different transmitters can use different channels for broadcasting. DTT and analog terrestrial TV (ATT) can use this method. The planning body is responsible for allocation of frequency for each transmitter. The power limit and the coverage should be planned in proper way, so that the secondary use is also facilitated well [16]. There must be a reference for the planning which is derived according to ITU recommendation. Regional conference decisions have also their own contribution in planning of national terrestrial TV [17, 18]. The planning includes fixing the outdoor receivers of antenna height 10 m and indoor mobile receivers to be 1.5 m. In Ref. planning configuration, reference values for receivable field strength, location probability and maximum interference level are set [6].

3.3 DTT protection

The primary transmission system must be kept from different interference sources which can result from white space use also [3]. The protection level for which the receiver of incumbent system kept unaffected is known to be protection ratio. It is the allowed level of signal quality determined by the deference of primary signals and interfering signals. It can be given as carrier to noise ratio, carrier to interference or interference to noise ratio [6]. It is limited by the national regulatory body. For Ethiopian case, since it is in region one, means its bandwidth is 8 MHz, the protection ratio is given by the following **Table 1**. **Table 1** below, illustrates the co-channel and adjacent channel protection ratio at 8 MHz channel bandwidth.

3.4 Propagation models

The frequency for terrestrial TV is different from radio frequency range which are used for other services. Although most of the propagation models developed and being modified are for mobile technology, there are also some propagation models which can cover the frequency range of terrestrial TV broadcasting.

ITU-R P.1546–5 [2, 3, 20]: This propagation model gives a point to area prediction of signals in the frequency range from 30 MHz to 3GHz with in a distance up to 1000 km and effective antenna height up 3000 m. It is statistical model for land, sea and mixed paths. The value of field strength is given in a graphical and tabulated form for some fixed values of frequency, effective antenna height and distance from the transmitter. To find the value of field strength with factors of different values from the givens, it is advisable to use interpolation (when the required value is between given values) or extrapolation (when the required value is out of the given values in the table or graph). The field strength interpolation (formulas for different parameters are given [3] as:

For distance, d, the interpolated field strength (E_d) can be obtained as

$$E_d = E_{inf} + \frac{(E_{sup} - E_{inf})\, log\left(d/d_{inf}\right)}{log\left(d_{sup}/d_{inf}\right)}\, dB\left(\mu\frac{V}{m}\right) \tag{1}$$

Protection ratio for 8 MHz [2, 19]	
Channel type to be protected	**Protection ratio (dB)**
Co-channel (ΔF = 0)	17
Adjacent channel (ΔF = ±1)	−36

Table 1.
Protection ratios [2, 3].

For frequency [3] *f, the interpolated field strength* (E_f) *can be obtained as*

$$E_f = E_{inf} + \frac{(E_{sup} - E_{inf})\, log\left(f/f_{inf}\right)}{log\left(f_{sup}/f_{inf}\right)}\, dB\left(\mu\frac{V}{m}\right) \tag{2}$$

and

For antenna height (h_1), the interpolated field strength E_{h_1} can be obtained as value [2],

$$E_{h_1} = E_{inf} + \frac{(E_{sup} - E_{inf})\, log\left(h_1/h_{inf}\right)}{log\left(h_{sup}/h_{inf}\right)}\, dB\left(\mu\frac{V}{m}\right) \tag{3}$$

Here, the subscripts *inf* and *sup* indicates the values of respective parameters given in the graph or table directly below and above the required value, respectively. In this model there are other factors like time percentage and location probability to be taken into account.

Okumura Hata Model [8]: Covers distance (d) up to 100 km. The operating frequency (f) ranges from 150 MHz up to 1.5GHz.

The loss for urban areas (L_{urban}) is given by:

$$L_{urban} = 69.55 + 26.16\, log\,(f) - 13.82\, log\,(h_t) - a(h_r) + (44.9 - 6.55\, log\,(h_t))\, log\,(d) \tag{4}$$

The correction factor $a(h_r)$ for middle and small cities is given by:

$$a(h_r) = (1.1\, log\,(f) - 0.7)h_r - (1.56\, log\,(f) - 0.8) \tag{5}$$

For open or rural areas (L_{rural}), it become:

$$L_{rural} = L_{urban} - 4.78(\,log\,(f))^2 + 18.33\, log\,(f) - 40.94 \tag{6}$$

Longley Rice/Irregular Terrain Model [21]: It is a model covers a frequency range from 20 MHz to 20GHz. The analysis uses several parameters and contains complex equations where it is simplified by a software called, Signal Propagation, Loss and Terrain Analysis Tool, SPLAT! It takes in the real terrain data and gives the output in coverage map or report in text format. Here we use point to point analysis, where the real data of the terrain is considered.

Hata Devidson Model [22]: The model uses frequency 30 MHz–1500 MHz, distance 1 km–300 km, Tx HAAT 20 m–2500 m and Rx antenna 1-10 m.

$$PL_{HD} = PL_{Hata} + A(h_1, d_{km}) - S_1(d_{km}) - S_2(h_1, d_{km}) - S_3(f_{MHz}) - S_4(f_{MHz}, d_{km}) \tag{7}$$

Where,

$$PL_{Hata} = 69.55 + 26.16\, log\,(f) - 13.82\, log\,(h_1) - a(h_2) + (44.9 - 6.55\, log\,(h_1))\, log\,(d) \tag{8}$$

and the correction factor for receiver antenna is the same as that for Hata model above. A and S_1 are factors that extend distance up to 300 km. S_2 is correction factor

Distance (km)	$A(h_1,d_{km})$	$S_1(d_{km})$
D_{km} < km	0	0
20 < d_{km} < 64.38	$0.62137(d_{km}-20)*[0.5 + 0.15log_{10}10(h_1/121.92)]$	0
64.38 < d_{km} < 300	$0.62137(d_{km}-20)*[0.5 + 0.15log_{10}10(h_1/121.92)]$	$0.174(d_{km}-64.38)$

Table 2.
Terms in the Hata Devidson model.

for HAAT to cover up to 2500 m. And S_3 and S_4 are correction factors for frequency to cover up to 1500 MHz. Hata Devidson Model terms are described in **Table 2** for different distance ranges.

And for other correction factors, $S_1(h_1,d_{km}) = 0.00784/log_{10}(9.98/d_{km})/(h_1-300)$, $h_1 > 300$ *m*, $S_3(f_{MHz}) = f_{MHz}/250log_{10}(1500/f_{MHz})$ *and* $S_4(f_{MHz},d_{km}) = [0.112log(1500/f_{MHz})](d_{km}-64.38)$, $d_{km} > 64.38$ *km.*

Egli Model [10]: A model for UHF and VHF in frequency range of 40 MHz–900-MHz. It calculates point to point link path loss for urban and rural as well. It is developed to include irregular terrains and its loss formula is given by:

$$PL_{Egli} = \begin{cases} 20\,log\,(f) + 40\,log\,(d) - 20\,log\,(h_t) - 10\,log\,(h_r) + 76.3, h_r \le 10m \\ 20\,log\,(f) + 40\,log\,(d) - 20\,log\,(h_t) - 10\,log\,(h_r) + 83.9, h_r > 10m \end{cases} \quad (9)$$

Where h_r is receiver antenna height AGL in meter, h_t is transmitter antenna height AGL in meter, d is the distance between transmitter and receiver in kilometers and f is the frequency in MHz.

4. System model and problem formulation

4.1 Problem formulation

For efficient utilization of free channels, keeping the interference minimum at the same time, the path loss model used by calculation engine must be efficient, where it should support both VHF and UHF. The path loss model should be less variant in different frequency ranges. If the path loss value has significant difference for frequency values in VHF and UHF, this will cause significance difference in contour coverage of the digital TV (DTV). This will on the other hand cause a variation in channels which are set free or occupied. Free channels for VHF may be occupied on UHF and vice versa. Propagation models selected for comparison must support the working frequency range of TV system. They must also be applicable for rural areas, since TV white space is assumed to be more applicable for rural areas. For our case, the TV frequency ranges from 174 MHz to 846 MHz. So, all the propagation models are defined for this frequency range. After the propagation model is selected, the coverage of the TV transmitter should be analyzed and this is used further to analyze the free channels which will be available as white space channels.

4.2 System Modeling

For the analysis purpose we have selected the above propagation models. These models are compared by their response for frequency variation. The analysis is done using Matlab software and SPLAT!. For those which can be simulated using Matlab, we have selected the transmitter to be Furi, which is located south west of Addis Ababa city and five different test points. To guarantee the reason of selection, Furi is working in frequencies under channel 7 and channel 42 [23]. The test points are selected in order to consider different points in geographical location, distance and terrain types. All the points shown in the **Figure 1** below are different in pathloss calculation parameters.

Frequencies of comparison are center frequencies for VHF and UHF frequency ranges of Ethiopian TV transmission band. The center frequency for VHF in Ethiopian case is taken by using the calculation formula:

$$f_{c(VHF)} = {(178MHz+220MHz)}/{2} \tag{10}$$

And its value is 199 MHz. For UHF case, the minimum frequency is 474 MHz and the maximum frequency is 786 MHz. The center frequency then is 636 MHz, with equation below.

$$f_{c(UHF)} = {(474MHz+786MHz)}/{2} \tag{11}$$

Other parameters taken as common for all models are:

- Transmitter height above ground level = 30 m, which results in HAAT of 639.42 m at Furi.

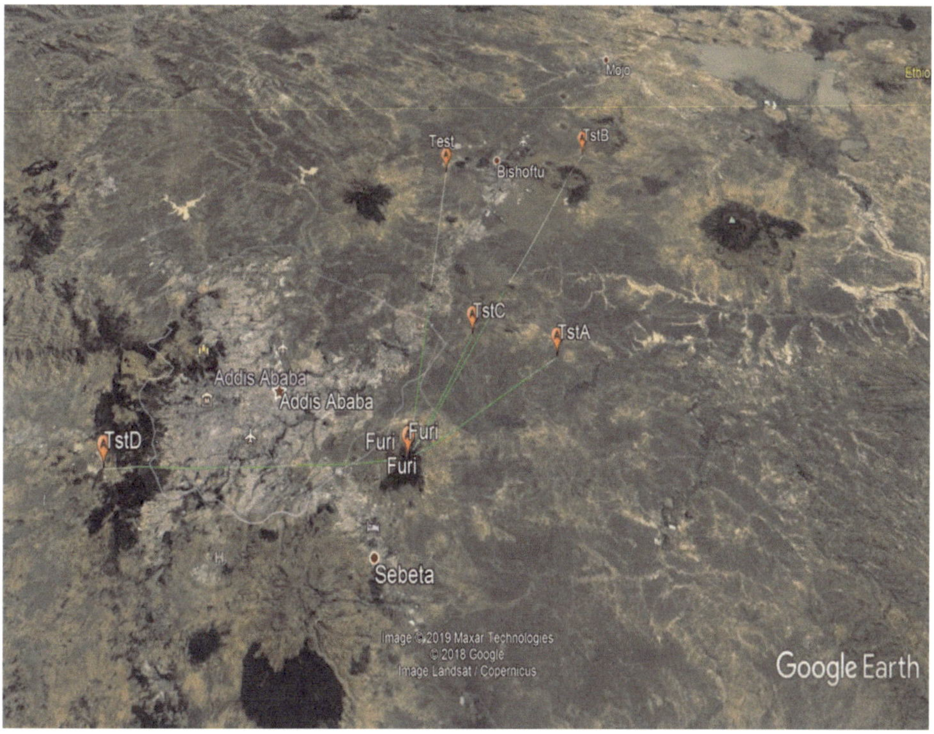

Figure 1.
Test points for pathloss analysis (adapted from Google Earth).

- Receiver antenna height = 1.5 m.
- Time and location probability for Longley Rice model are 0.95 and 0.5 respectively
- And the distance between Tx. And Rx. is 35 km (from **Figure 1**)

For the pathloss analysis of Longley rice model the real data is used for different locations. These test points (shown in **Figure 1**) are selected at random. The points selected, although they are random, are selected from different locations to include different propagation path types.

5. Discussion of results

The response of propagation models for frequency variation from VHF and UHF shows different value for various models. For the models described above, a detailed comparison is made so as identify which propagation model is best fit for quantifying the available TVWS channels in Ethiopia's TV frequency interest. When the proposed propagation model (ITU-R P.1546–5) is implemented after interpolation, the pathloss must be changed toa new loss with the following formula:

$$\mathrm{L} = 139.3\text{-}E + 20log(f)\ where\ E\ is\ in\ dB\left(\mu\frac{V}{m}\right) where\, f\ in\ MHz \tag{12}$$

After this transformation is made different calculations are performed and compared with the loss difference of other models by using the scenario depicted in **Figure 1**.

As can be seen from **Figure 2**, the comparison between different pathloss models for VHF and UHF TV transmission spectrums are illustrated. The path loss is computed based on the scenario illustrated in **Figure 1**. As seen from the **Figure 1**, the proposed propagation model (ITU-R.P.1546–5) gave a much better result than others. Hence, so as to compute the available TVWS free spectrums at a certain place and location, ITU-R.P.1546–5 is the best propagation model.

Based on the results obtained in **Figure 2**, sensitivity of the propagation models comparison can be best described at different frequency ranges as depicted in **Figure 3**. As seen in **Figure 3**, the Longley Rice model is very sensitive at different frequency and distance ranges. It has also varying value of pathloss differences at different test points. So, taking the average of these values, we have a difference of *18.28.* The other values, as can be seen from the graph, are somewhat with constant value. Comparing these values, the pathloss model, ITU-R P.1546–5 shows the minimum variation for frequency variation of TV signals from VHF to UHF with a value of 9.9736. The other model with the nearest value is Okumura Hata model. But selecting ITU-R P.1546–5 model has better advantage in consideration of terrain data. So, ITU-R P.1546–5 is the best pathloss model to represent TV signal coverage and signal modeling. Using this pathloss model, the free channels found free with assumption that the WSD is placed at the center of each TV transmitter, which is the worst-case scenario for WSD placement, we have found free channels as shown in **Figure 4** below.

As can be seen from **Figure 4**, among the 58 TV transmitter channels, sites which have 1 occupied co-channel and adjacent channel is 28% and 47% respectively.

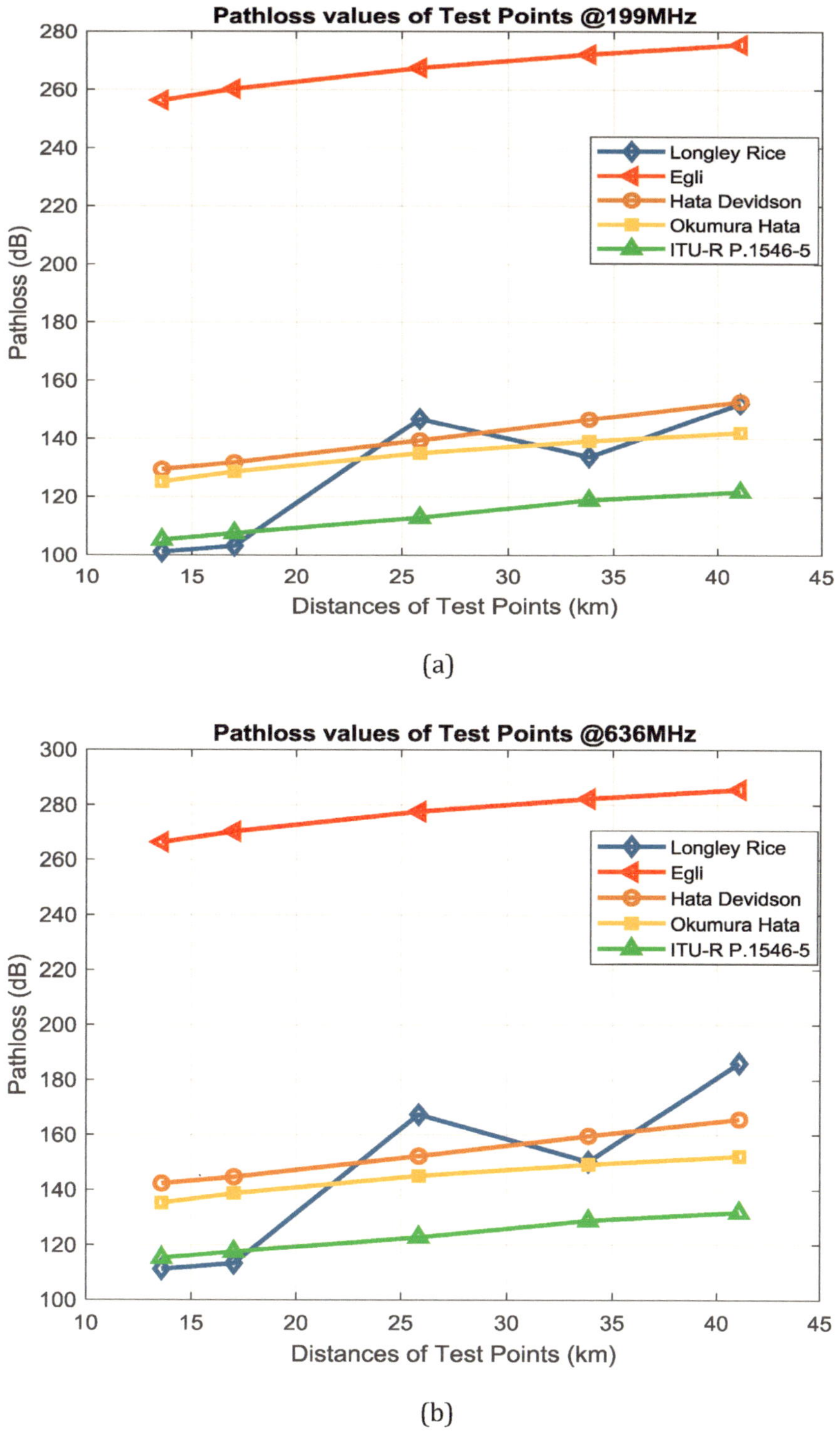

Figure 2.
Pathloss model results (a) and (b).

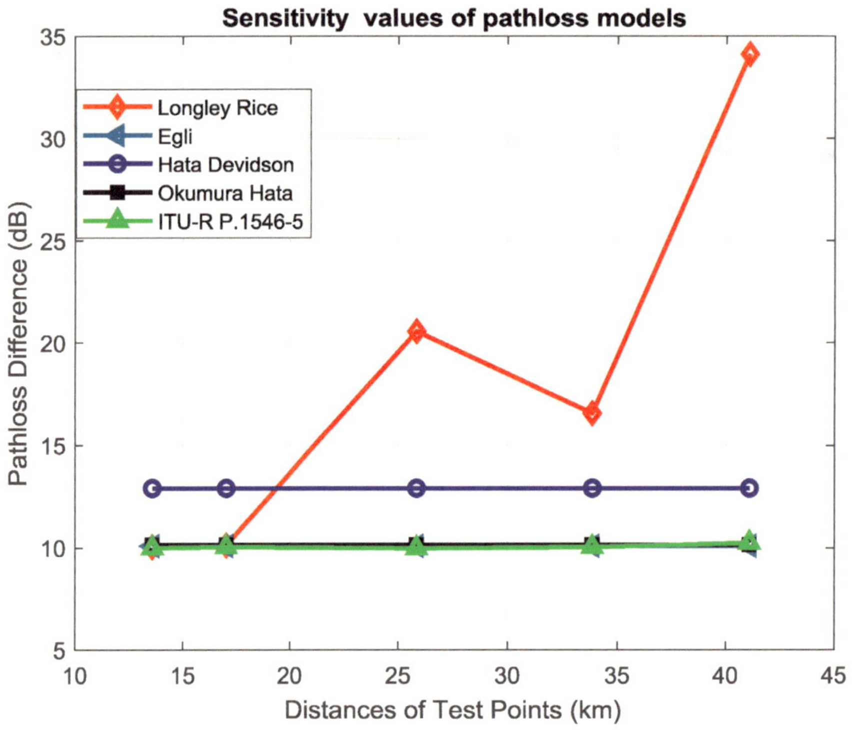

Figure 3.
Variations of Pathlosses for VHF and UHF.

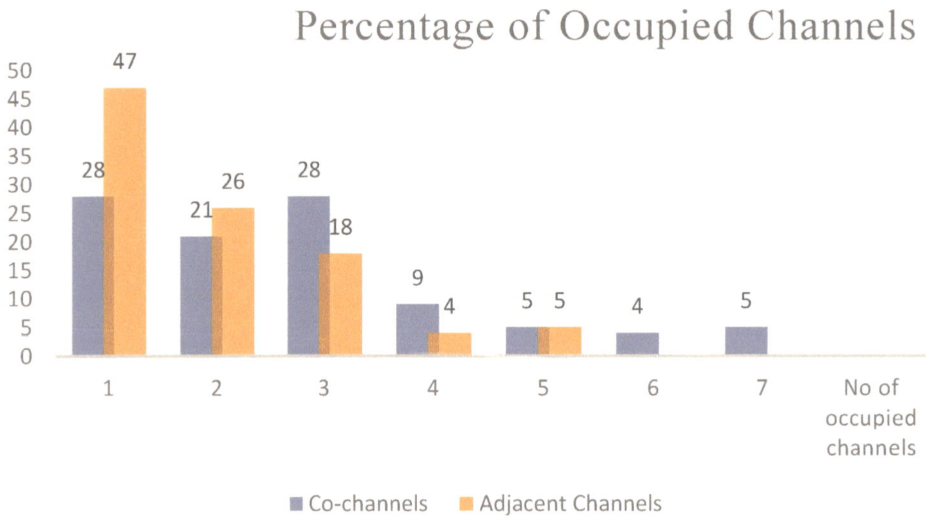

Figure 4.
Number of occupied channels and their respective percentages for Co and adjacent channels.

Likewise, sites which have 2 occupied co-channels and adjacent channels is 21% and 26% respectively and so on up to 7 occupied co-channels and adjacent channels. These values indicated that there are 51 to 57 free unoccupied free channels in Ethiopia among the 58 TV transmitter. Exploiting such vacant TV spectrums for affordable wireless broadband, machine to machine communication, vehicle to vehicle communication vision sensors and etc. are very promising. **Figure 4** also depicts that the highest capacity for secondary use is concentrated around one up to three. This implies that the majority of free channels, which can reach 80% of free channels, have a minimum of 51 channels (contiguous bandwidth of 408 MHz) available.

6. Conclusion

From the given models, looking at the values of differences in loss at different frequencies, small variations are observed at ITU-R P.1546–5 model. Hence, we can select ITU-R P.1546–5 recommendation to be less sensitive to the variations in frequency from VHF to UHF. So, it is the best model to represent the TV signal with less variation in loss value whether the TV is operating in VHF or UHF band by incorporating the real terrain data of Ethiopia. Using this pathloss model, we have calculated free channels and found many free channels. Out of 58 TV channels, a minimum of 51 channels are free for secondary use. It indicates there is a sufficient amount of contiguous bandwidth up to 408 MHz for white space use.

Author details

Habib Mohammed[1]*, Tessema T. Terefe[2] and Sultan Feisso[2]

1 School of Electrical and Computer Engineering, Addis Ababa Institute of Technology (AAiT), Addis Ababa University, Addis Ababa, Ethiopia

2 Department of Electrical and Computer Engineering, College of Electrical and Mechanical Engineering, Addis Ababa Science and Technology University, Addis Ababa, Ethiopia

*Address all correspondence to: habib.mohammed@aait.edu.et

References

[1] J. S. Seybold, Introduction to RF propagation, Hoboken, New Jersey: John Wiley & Sons, 2005.

[2] H. Mohammed, K. Katzis, L. Mfupe, E. Teshale,"Practical Implementation of Geo-location TVWS Database for Ethiopia, "ICAST2020-8th EAI International Conference on Advancements of Science and Technology, ICAST2020.[in press] press]

[3] Habib M. Hussien, K. Katzis, L. P. Mfupe, and E. T. Bekele, "Coexistence of TV White Space Devices and DTV Services in Ethiopian Geolocation White Space Spectrum Database," in 2019 IEEE 24th International Workshop on Computer Aided Modeling and Design of Communication Links and Networks (CAMAD), 2019, pp. 1–5.

[4] K. Katzis, L. Mfupe and H. M. Hussien, "Opportunities and Challenges of Bridging the Digital Divide using 5G enabled High Altitude Platforms and TVWS spectrum," *2020 IEEE Eighth International Conference on Communications and Networking (ComNet)*, Hammamet, Tunisia, 2020, pp. 1-7, doi: 10.1109/ComNet47917.2020.9306090.

[5] Tessema T. Terefe, Habib M. Hussien, Sultan F. Meko, "Reserved Distance Determination in Incumbent and TV White Space System Coexistence", ICAST2020.[in press]

[6] L. Mfupe, F. Mekuria, M. Mzyece, "Geo-location White Space Spectrum Databases: Models and Design of South Africa's First Dynamic Spectrum Access Coexistence Manager," KSII TRANSACTIONS ON INTERNET AND INFORMATION SYSTEMS, vol. 8, no. 11, Nov. 2014.

[7] M. Mekonnen, "Feasibility Study of TV White Space for Broadband Internet Services: The Case of Rural Ethiopia," *A Thesis Submitted to the School of Graduate Studies of Addis Ababa University in Partial Fulfillment of the Requirements for the Degree of Masters of Science in* Electrical Engineering, January, 2017.

[8] K. Stylianos, "Modelling and Coverage Improvement of DVB-T Networks," *A thesis submitted for the degree of Doctor of* Philosophy, March 2018.

[9] A. E. S. C. J. O. R. Gorrepati, "Improved Estimation of TV White Spaces in India using Terrain Data," Twenty-third National Conference on Communications (NCC), 2017.

[10] A. M. Fanan, N. G. Riley, M. Mehdawi, O. Alfahad, "Performance of a TV white space database with different terrain resolutions and propagation models," Telfor Journal, vol. 9, no. 2, 2017.

[11] S. Kasampalis, P. I. Lazaridis, Z. D. Zaharis, A. Bizopoulos, S. Zettas, J. Cosmas, "Comparison of Longley-Rice, ITU-R P.1546 and Hata-Davidson propagation models for DVB-T coverage prediction," *IEEE BMSB,* July 2014.

[12] T. S. Rappaport, Wireless Communications, Principles and Practice, Upper Saddle River, NJ: Prentice-Hall, , 2002.

[13] M. Shahajahan, A. Q. M. Abdulla Hes-Shafi, "Analysis of Propagation Models for WiMAX at 3.5 GHz," *This thesis is presented as part of Degree of Master of Science in* Electrical Engineering, September 2009.

[14] ITU-R, "Recommendation ITU-R P.1411-8, Propagation data and prediction methods for the planning of

short-range outdoor radio communication systems and radio local area networks in the frequency range 300 MHz to 100 GHz," July 2015.

[15] P. G. Brown, K. Tsioumparakis, M. Jordan, and A. Chong, "UK Planning Model for Digital Terrestrial Television Coverage," R&D White Paper, WHP 048, BBC, September 22, 2002.

[16] GE-06:, "Final Acts of the Regional Radiocommunication Conference for planning of the digita," Vols. ITU,RRC-06.

[17] "Final Acts of Europe VHF/UHF Broadcasting Conference," ITU, 1961.

[18] "The Chester 1997 Multilateral Coordination Agreement relating to Technical Criteria, Coordination Principles and Procedures for the Introduction of Terrestrial Digital Video Broadcasting (DVB-T)," CEPT, Chester, July 25, 1997.

[19] Ofcom, UK, "TV white spaces: Approach to co-existence," *Technical report,* September 4, 2013.

[20] ITU, "ITU-R P.1546-5, Method for point-to-area predictions for terrestrial services in the frequency range 30 MHz to 3 000 MHz," P Series, Radiowave propagation, Geneva, 09/2013.

[21] A. G LONGLEY, P. L. RICE, "Prediction 3f Tropospheric Rado Transmission Loss Over Irregular Terrain A Computer Method," National Technical Information Service, U. S. Department Of Commerce, 1968.

[22] Hata, Davidson, "A Report on Technology Independent Methodology for the Modeling, Simulation and Empirical Verification of Wireless Communications System Performance in Noise and Interference Limited Systems Operating on Frequencies between 30 and 1500MHz," *TIA TR8 Working Group, IEEE Vehicular Technology Society Propagation Committee*, May 1997.

[23] "Roadmap for the transition fRom analogue to digital terrestrial television in Ethiopia," ITU, April 2012.

Chapter 6

Array Antenna for Reconfigurations

Smrity Dwivedi

Abstract

Reconfiguration array antenna is usually achieved in terms of pattern, frequency, and polarization by changing the relative amplitudes, phases of the excitation distribution, and material characteristics present in the array, at the cost of simple and complex feeding networks both. In this chapter, basically different types of array antennas are discussed with its design with reconfigurations. Also, for high generation communication system, a widely used graphene material is taken to get reconfiguration in antenna using array form. There are so many other techniques being used and discussed to get reconfiguration. To get reconfiguration at high frequency, i.e., terahertz frequency, graphene is the best material to use in antenna to solve many applications nowadays compared with only few materials, which is useful design. Here, few designs have been introduced to understand the reconfiguration in array antenna for broadband applications. The proposed chapter represents advancement on the development of multipurpose antennas having different applications in communication, broad banding, detection, etc., in recent research systems.

Keywords: array antenna, reconfigurable techniques, frequency reconfiguration, pattern reconfiguration, polarization reconfiguration, graphene, liquid crystal

1. Introduction

A reconfigurable antenna is an antenna having a capability to modify its properties dynamically, in a controlled and reversible manner. The need for multifunctional (e.g., direction finding, beem steering, radar, control, and command), high-performance and cost-effective devices within a confined volume places a greater burden on today's transmitting and receiving systems. Reconfigurable antennas are a solution to this problem. Reconfiguring an antenna is achieved through deliberately changing its frequency, polarization, or radiation characteristics. Many techniques are there to achieve this change by redistributing antenna currents and thus altering the electromagnetic fields of the antenna's effective aperture, thereby adapting to changes in environmental conditions or system requirements (i.e., enhanced bandwidth, changes in operating frequency, polarization, and radiation pattern) [1, 2]. This concept can significantly reduce the number of components, hardware complexity, and cost of the communication systems. The first patent on reconfigurable antennas appeared in 1983 by Schaubert [3]. Excellent overview of reconfigurable antennas, with many examples, is given in [4].

The type of reconfiguration and the technique to achieve need to be addressed before designing an antenna. There are four fundamental reconfigurable properties of antenna, i.e., frequency of operation, radiation pattern, polarization, or a combination of any of these properties [4].

These antennas are those antenna systems that are capable of modifying their properties such as frequency, pattern, and polarization dynamically, within a controlled strategy, and with permissible reversibility. The reconfigurability of the antenna array is based on the modification of the geometry of system or behavior of its elements with regular maintenance of the efficiency of the antenna against changes on its environment and mission objectives. Reconfigurable antennas are usually result in low cost, lightweight, low volume but with complexity of maintenance and proper repairing due to being able to offer the same functionalities as multiple conventional antennas [5]. Moreover, the modification of the antenna elements in terms of electrical characteristics causes strong complexity in design of feeding networks results [6–8], and it is more worse if extra controlling elements are added. Due to this reason, the use of arrays technology by introducing parasitic elements in antenna is very common. Because of these parasitic elements, extra current is induced due to near-field effects without any need of a feeding network in its design, creating beam pattern reconfiguration with a noticeable very low complexity [9, 10]. The most common antennas that use parasitic elements based on linear dipole arrays are known as Yagi-Uda antenna arrays [11]. In such antenna geometry, one more element is added along with parasitic linear dipole known as driven, used in front of those to get reconfiguration in radiation pattern. Also, linear and planar arrays are being used to achieve same goal by use of switch on and off condition, one by one in the same elements [9]. One more antenna that is printed dipole antenna having single parasitic element is used for wide scanning at the cost of little deviation in the gain of antenna, known as beam scanning antenna [12]. As of interest in beam pattern reconfiguration for the antenna, few dielectric materials can also be responsible for reconfiguration by changing its value due to environment behavior and surrounding, giving impressive results for so many applications. As relative dielectric constant of medium, which is gaseous, particularly air is the main ingredient having constant near unity, is basically changed due to change in temperature, pressure, and humidity that can easily be checked during hygrometry measurements using Yagi-Uda antennas [13–15]. So, the environment effect on dielectric constant is widely used to achieve reconfiguration. Later, this constant can also be changed in the presence of airborne particles and due to pollution in industrial and urban area such as droplets in clouds, which is relevant in today's antenna design [16]. One more aspect is important during the design of reconfigurable array antenna, which is dynamic range ratio that should be controlled in the amplitude of excitation, which is one main constraint. This dynamic range ratio is the ratio between the maximum and the minimum excitation amplitude of the array elements in antenna, which allows the practical realization of feeding networks, under which noncomplex design is offered with less number of power dividers or the design of micro stripline should become simple. There are so many methods given to work together with pattern and feeding network, which are mainly to reduce the dynamic range ratio, specified value for each excitation amplitude, or controlling phase. Reconfigurability in array antennas is second significant capability that may be requested nowadays from antenna designer to get wide applications. In current scenario, communication systems might in fact have to accomplish multitasking missions, in which the pattern must be reshaped by keeping the constant the value of excitation amplitudes of the elements and also modifying the sole excitation phases. Getting frequency reconfiguration and polarization reconfiguration without changing the parameters of antenna is also very promising and difficult task.

2. Reconfiguration

A reconfigurable antenna is an antenna that is capable of changing frequencies and other characteristics as per requirement. It can be used by using diodes as switches. It is equipped for altering its recurrence and radiation properties progressively, in a controlled and reversible way. So as to give a unique reaction, reconfigurable radio wires incorporate an inward component (for example, RF switches, varactors, mechanical actuators, or tunable materials) that empowers the purposeful redistribution of the RF flows over the receiving wire surface and produces reversible alterations of its properties. Reconfigurable radio wires contrast from savvy reception apparatuses in light of the fact that the reconfiguration component lies inside the receiving wire, and it is opposed to in an outer beam forming system. The reconfiguration ability of reconfigurable receiving wires is utilized to augment the reception apparatus execution in a changing situation or to fulfill changing working prerequisites.

2.1 Types of reconfiguration

Reconfigurable reception apparatuses can be grouped by the radio wire parameter that is powerfully balanced, commonly the recurrence of activity, radiation design, or polarization. Different types of reconfiguration are mentioned as given.

2.1.1 Frequency reconfiguration

Recurrence reconfigurable radio wires can modify their recurrence of activity powerfully. They are especially helpful in circumstances where a few interchanges frameworks merge in light of the fact that the different receiving wires required can be supplanted by a solitary reconfigurable radio wire. Recurrence reconfiguration is by and large accomplished by physical or electrical adjustments to the reception apparatus measurements utilizing RF switches, impedance loading, or tunable materials.

2.1.2 Radiation reconfiguration

Radiation design reconfigurability depends on the purposeful change of the circular circulation of the radiation design. Bar guiding is the most expanded application and comprises of controlling the heading of greatest radiation to boost the receiving wire gain in a connection with cell phones. For example, reconfigurable receiving wires are typically planned utilizing mobile/rotatable structures or switchable and responsively stacked parasitic elements. Over the most recent 10 years, metamaterial-based reconfigurable reception apparatuses have picked up consideration due to their little structure factor, wide shaft guiding extent, and remote applications. Plasma radio wires have additionally been explored as options with tunable directivities.

2.1.3 Polarization reconfiguration

Polarization reconfigurable radio wires are fit for exchanging between various polarizations modes. The ability of exchanging between flat, vertical, and roundabout polarizations can be utilized to decrease polarization bungle misfortunes in versatile gadgets. Polarization reconfigurability can be given by changing the harmony between the various methods of a multimode structure.

2.1.4 Compound reconfiguration

Compound reconfiguration is the capacity of all types of reconfigurations while tuning a few receiving wire parameters, for example, recurrence and radiation design. The most widely recognized use of compound reconfiguration is the mix of recurrence spryness and bar checking to give improved ghostly efficiencies. Compound reconfigurability is accomplished by consolidating in a similar structure distinctive single parameter reconfiguration techniques or by reshaping progressively a pixel surface.

2.2 Techniques for reconfiguration

Four major techniques are available to implement reconfigurable antennas, i.e., electrically reconfigurable antennas, optically reconfigurable antennas, physically reconfigurable antennas, and materially reconfigurable antennas as shown in **Figure 1**. Electrically reconfigurable antennas use electrical switches such as RF-MEMS, PIN diodes, or varactors to redistribute the surface current and alter the antenna radiating structure topology and/or radiating edges [17–20]. Optically reconfigurable antennas rely on photoconductive switching elements. Physically reconfigurable antennas can be achieved by altering the structure of the antenna [21, 22]. Finally, reconfigurable antennas can also be implemented through tunable materials such as liquid crystals and graphene as in **Figure 1** [4, 23].

2.3 Comparison between different reconfiguration techniques

Electronic switching components have been widely used to reconfigure antennas, especially after the appearance of RF-MEMS in 1998. One of the main advantages of such components is their good isolation and low-loss property. While RF-MEMS represents an innovative switching mechanism, its response is slower than PIN diodes and varactors, which have a response on the order of nanoseconds. The ease of integration of such switches into the antenna structure is matched by their nonlinearity effects (capacitive and resistive) and their need for high voltage (RF-MEMS, varactors). The activation of such switches requires biasing lines that may negatively affect the antenna radiation pattern and add more losses. The incorporation of switches increases the complexity of the antenna structure because of the need of additional bypass capacitors and inductors, which increase the power consumption of the whole system.

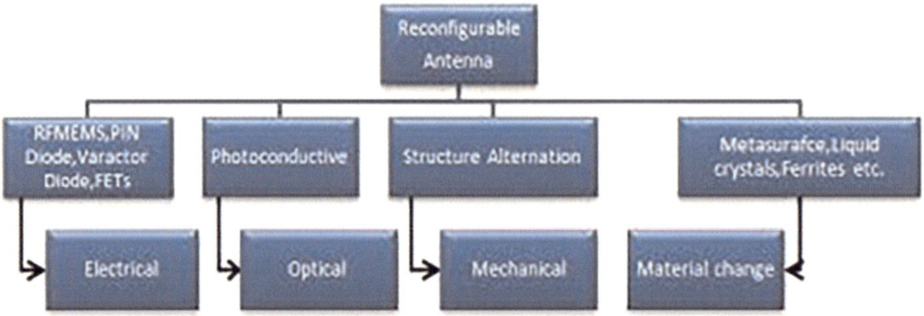

Figure 1.
Antenna reconfiguration techniques [4].

Even though optical switches are not much famous, they offer a reliable reconfiguration mechanism especially in comparison to RF-MEMS. The activation or deactivation of the photoconductive switch by shining light from the laser diode does not produce harmonics and intermodulation distortion due to their linear behavior. Moreover, these switches are integrated into the antenna structure without any complicated biasing lines, which eliminates unwanted interference, losses, and radiation pattern distortion. Despite all these advantages, optical switches exhibit lossy behavior and require a complex activation mechanism. **Table 1** shows a comparison of the characteristics for the different switching techniques used on electrically and optically reconfigurable. The advantages of using physical reconfiguration techniques lie in the fact that they do not require bias lines or resort to laser diodes or optical fibers. However, their disadvantages include slow response, cost, size, power source requirements, and the complex integration of the reconfiguring element into the antenna structure. As for the materially reconfigurable antennas, one major advantage of using liquid crystals is their moderate losses and ease of tuning. Graphene is one other tunable wonder material, which exhibits excellent properties. At submillimeter wave where the footprint of antenna is very small and integrating switches is not possible; we can use these tunable materials to make reconfigurable antennas.

Reconfiguration is possible at GHz as well as THz frequencies as per requirement and applications. Nowadays, advance communication in 5G or more than 5G requires high frequency, which can give high bandwidth for wide range uses. In **Tables 2** and **3**, lists of reconfigurable antennas are given, which are already designed and in use.

The state of the art of research on material-based reconfigurable antennas at GHz frequencies from **Table 2** shows that only few LC materials are explored to achieve reconfiguration. At microwave frequency, graphene is very less used due to high loss. Even LC materials show moderate losses at microwave frequency but because of fabrication limitations, LC is preferred over graphene at microwave frequency. Graphene is much more used than LC at THz frequencies due to unique effect of surface plasmon polariton (SPP) wave. Frequency reconfiguration range depends on the range of permittivity achieved with different LC materials. The designed antenna is based on K15 with no reported gain in literature as given -1 to 0 dBi [24, 25, 29]. A novel design of microstrip patch antenna is proposed based on K15 LC [28]. This antenna has 6.2 dB of gain with good impedance matching. The designed antenna is based on E7 LC reported in [26]. The reported gain is 0.7–1.1 dBi, whereas the gain is not reported [27]. The designed antenna based on GT3-23001 LC is proposed in [30] with the designed structure of 1×4 element antenna array with 12 dBi gain. The antenna applies a double-layer structure with no reported gain [31]. The designed

Electrical property	RF MEMS	PIN diode	Optical switch
Voltage [V]	20–100	3–5	1.8–1.9
Current [mA]	0.01–0.05	3–20	0–87
Power consumption [mW]	0.05–0.1	5–100	0–50
Switching speed	1–200 μs	1–100 ns	3–9 μs
Isolation (1–10 GHz)	Very high	High	High
Loss (1–10 GHz) [dB]	0.05–0.2	0.3–1.2	0.5–1.5

Table 1.
Electrical properties of electrically and optically switching [4].

Ref	Year/journal	Frequency	Reconfiguration technique	Reconfiguration property	Reconfiguration range
Missaoui et al. [24]	2011 JCEMS	5 GHz	K15 LC	Frequency	4.5–5.5 GHz#
Papanicolaou et al. [25]	2015 EL	4.77 GHz	K15 LC	Frequency	4.64–4.77 GHz#
Polycarpou et al. [26]	2014 TAP	5.8 GHZ	E7 LC	Frequency	5.45–5.84 GHz#
Liu et al. [27]	2008 EL	5 GHz	E7 LC	Frequency	5.3–5.75 GHz#
Missaoui et al. [28]	2016 DIPED	2.53 GHz	K15 LC	Frequency	2–2.53 GHz#
Martin et al. [29]	2003 EMC	5 GHz	K15 LC	Frequency	4.94–5.07 GHz#
Zhao et al. [30]	2017 JPHOT	15.3 GHz	GT3–23001 LC	Frequency and Radiation Pattern	14.5–16.4 GHz, −20° to 20°*
Pan et al. [31]	2018 ICTCE	6.02 GHz	GT3–23001 LC	Frequency	5.91–6.02 GHz#
Sethi et al. [32]	2016 APMC	6.6 GHz	DLA 100–100	Frequency	6.4–6.6 GHz#
Yasir et al. [33]	2017 LAWP	5 GHz	Graphene	Frequency	4.50–5.05 GHz#
Leng et al. [34]	2015 APSURSI	1.588 GHz	Graphene	Frequency	0.83–1.588 GHz*

#Simulated and measured result.
**Simulated result.*

Table 2.
Comparison of reconfigurable antennas at GHz frequency reported in the literature.

antenna is based on transparent DLA 100-100 LC with poor impedance matching as reported in [32]. Gain of this structure is not reported. The designed antenna is based on graphene with gain 0.76–2.38 dB in [32] and 0.1–1.9 dB in [33, 34]. At microwave frequency, biased graphene works as a lossy metal, and hence, gain of the antenna reduces due to high power dissipation and increased sheet resistance. So, based on the above literature review, LC is chosen over graphene for the proposed S (2–4 GHz) band antenna for WLAN/WiMAX applications. Comparative analysis of antenna performance and tuning range is done with different LC materials.

The state of the art of research in material-based reconfigurable antennas at THz frequencies from **Table 3** shows that at higher frequency, graphene is preferred over LC. Property of spp waves in graphene help in extreme miniaturization of the antenna size with good performance. Most of the antennas designed in 0–1 THz band are given in [4, 35–43, 45]. The graphene-based microstrip patch antenna is designed on silicon substrate with gain 2.43–4.19 dB, efficiency 48–51%, and return loss 13–26 dB [35]. The graphene-based microstrip patch antenna designed on glass (SiO_2) substrate with capacitive-coupled transmission line is given in [36]. Reported gain is 5.08 dB, efficiency is 66%, and return loss is 39.19 dB. The graphene-based patch antenna is designed on silicon substrate with efficiency 16–29%, return loss 23–30 dB, and no reported gain [37]. The graphene-based microstrip patch antenna is designed on polyimide substrate with gain 4.93–5.07 dB, return loss up to 35 dB, and efficiency 86.53–6.85% reported in [4]. The graphene-based Yagi-Uda antenna is designed on glass (SiO_2) substrate with gain 6.5 dB and return loss 19 dB [38]. The graphene-based patch antenna designed on glass (SiO_2) substrate with directivity 2.99–5.56 dB and return loss 24–34 dB is given in [39]. The graphene-based 1×2 microstrip patch antenna array designed on glass (SiO_2) substrate with efficiency 8–43%, return loss 16–30 dB, and directivity 9 dB is given in [40]. The graphene-based patch antenna is

Ref	Year/ journal	Frequency	Reconfiguration technique	Reconfiguration property	Reconfiguration range
Yi Huang et al. [35]	2012 TNANO	0.86 THz	Graphene	Radiation pattern	±30°*
Goyal et al. [36]	2018 MOTL	1.0 THz	Graphene (0.09–0.45 eV)	Frequency	0.96–1.03 THz*
Dash et al. [37]	2017 iAIM	1.24 THz	Graphene (0.5–0.8 eV)	Frequency	0.98–1.24 THz*
Anand et al. [38]	2014 ijleo	0.775 THz	Graphene	Frequency	0.725–0.775 THz*
Wu et al. [39]	2016 EuCAP	1.88 THz	Graphene (0–0.4 eV)	Radiation pattern	2 beam pattern*
Dash et al. [40]	2018 IET-MAP	2.56 THz	Graphene (0.3–0.6 eV)	Frequency	2.56–4.98 THz*
Moradi et al. [41]	2018 ijleo	0.51 THz	Graphene (0.3–0.7)	Frequency	0.505–0.52 THz*
Tamagnone et al. [41]	2012 Appl. Phys. Lett.	1.8 THz	Graphene (0–0.2 eV)	Frequency	0.8–1.8 THz*
Hlali et al. [42]	2018 PIER C	2.5 and 5 THz	Graphene (0.1–0.4 eV)	Frequency	2.5–2.95 THz, 5.06–5.84 THz*
Zhang et al. [43]	2018 APCAP	1 THz	Graphene (0.1–2 eV)	Frequency and radiation pattern	1–2 THz*
Palomino et al. [44]	2015 TAP	1 THz	GT3-23001	Radiation pattern	55°#

#*Simulated and measured result.*
**Simulated result.*

Table 3.
Comparison reconfigurable antennas at THz frequency reported in the literature.

designed on GaAs substrate with efficiency 20% and no reported gain [41]. The graphene-based microstrip patch antenna is designed on Duroid substrate with return loss up to 40 dB and no reported gain [45]. The graphene-based square spiral antenna is designed on quartz substrate with return loss up to 37 dB, impedance bandwidth 26.68%, and no reported gain [42]. The LC-based reflect array (with 54 × 52 multiresonant cells) in which LC is inserted between quartz and silicon is designed on FR-4 with gain 19.4 dB and return loss 13 dB [43]. So, based on the literature review shown in **Table 3**, it is summarized that graphene-based antennas are much more suitable than LC at THz frequencies. Most of the graphene antennas proposed till now are based on either thick layer of graphene or multilayer graphene with moderate performance. Mostly frequency reconfiguration is reported, very few on radiation pattern, and almost none on polarization reconfiguration. The proposed antenna array is based on monolayer graphene with frequency, polarization, and radiation pattern reconfiguration all together with good performance.

3. Design schematic of array antenna for reconfiguration for different materials

Here, two different materials are being used to get reconfigurability in the array antenna, which is basically liquid crystal and graphene. Both materials are having properties to achieve high gain compatible with gigahertz as well as terahertz frequency. Liquid crystal gives good response on gigahertz frequency, whereas graphene gives high performance on terahertz frequency.

3.1 Using liquid crystal and graphene

Liquid crystal molecules show an additional state of matter, which lies between a liquid and a solid state [44, 46]. Similarly graphene also features surface plasmon polariton wave at terahertz frequency. Due to the excellent properties and ease of tuning, these materials have become the topic of great interest in the field of reconfigurable antennas. Fundamental properties of liquid crystals and graphene are presented in the first and second sections of this chapter, respectively.

3.1.1 Fundamental of liquid crystal

The most common states of matter are solid, liquid, and gas depending on the temperature and pressure. In the solid state, the molecules are strongly bonded by intermolecular forces either in a regular order, so-called crystalline solid, or in an irregular order, so-called amorphous solid. The strengths of the intermolecular forces in the LC materials are not homogeneous in all directions because the material is formed by anisotropic molecules. With the increase of temperature, molecules vibrate excessively and breakdown the weak intermolecular bonds, leading to a drop off in the positional order. This middle state is defined as the liquid crystal state, where the molecules exhibit orientational order like that of a solid crystalline but can still flow like a liquid. **Figure 2** shows the schematic representation of a nematic liquid crystal. In an LC bulk, the molecules are aligned parallel to their long axis because of the shape anisotropy. In the macroscopic scale, the time averaged direction of the director is along z axis, which is along the long axis of the molecules. In the both solid and nematic phases, a director can be assigned, whereas the level of molecular ordering is different. This is quantitatively defined by an order parameter S.

3.1.2 Fundamentals of graphene

The recent discovery of graphene atomically thin layer of graphite by Novoselov, Geim et al. in 2004 brought a period of scientific and technological research. It's been only 15 years to the history of graphene, and it has already attracted considerable attention in fields ranging from material science, nanotechnology to physics, electrical and electronics devices, and circuit applications beyond what is possible with today's silicon chips. Graphene, an extremely good conductor of electricity, comes as an alternative material replacing extensively used materials such as "silicon" or "metals" in field of electronics, in order to further shrink the device size. So, we can clearly say that graphene represents a conceptually new class of materials that are only one atom thick and, on this basis, offers new inroads into low-dimensional physics that has never ceased to surprise and continues to provide a fertile ground for applications. This section reviews the basic theoretical aspects of graphene material and discusses the desirable electromagnetic and mechanical properties possessed by graphene that would assist in providing flexible and reconfigurable antenna structures.

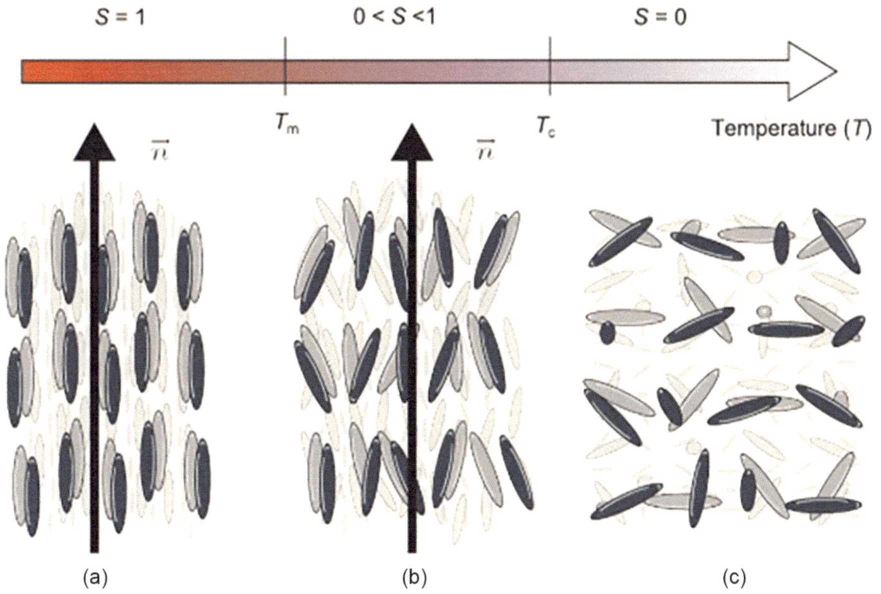

Figure 2.
Schematic of a nematic LC material depending on the temperature: (a) crystalline solid state with the order parameter S = 1. (b) Nematic phase, where $\vec{n}$ indicates the time average directions of the molecules. (c) Liquid phase with lowest molecular order [47].

3.1.3 Reconfigurable array antenna using liquid crystal material

The antenna design procedure consists of three steps. First, an inset fed patch antenna is designed with copper patch printed on 1.5 mm thick RT/Duroid 5880 substrate with permittivity ε_r = 2.2, loss tangent = 0.0009. To decide the dimensions of the inset feed patch antenna, we have to first specify the operating resonant frequency f_r, the height of the substrate h, and the permittivity of the dielectric substrate material ε_r (**Figure 3**).

Now LC material is placed between the layers of patch and RT/Duroid 5880 substrate. Once LC material is inserted, the effective permittivity of substrate will have an effect on both RT/Duroid 5880 and LC material. The orthographic view of the antenna is shown in **Figure 4**.

The antenna is designed by means of the simulator ANSYS HFSS. ANSYS HFSS is a 3D full-wave EM solver. This commercial software uses finite element method for solving electromagnetic structures. The feeding structure of the antenna consists of inset fed microstrip transmission line with 50 Ω impedance. To incorporate the array, corporate feed is used.

Figures 5 and **6** show the return loss S_{11} plotted as a function of frequency in the range of 2–3 GHz demonstrating the impedance matching conditions for patch antenna array without LC and with LC material. Here, frequency reconfiguration has been achieved. It can be shown that the return loss of the patch antenna array without LC reaches the maximum value of 22.95 dB at resonating frequency 2.45 GHz. After inserting the LC between substrate and patch, the resonating frequency shifts to the small amount and return loss reaches the value of 12.60 dB. Hence, a 1 × 2 antenna array is designed in this work and the comparative investigation of the antenna

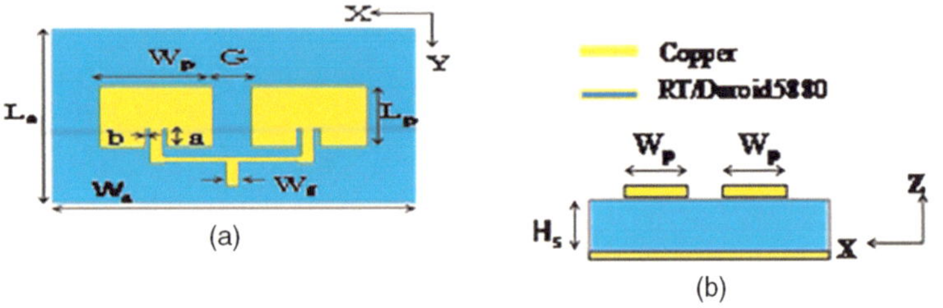

Figure 3.
(a) Top view of 1 × 2 antenna array and (b) front view of 1 × 2 antenna array.

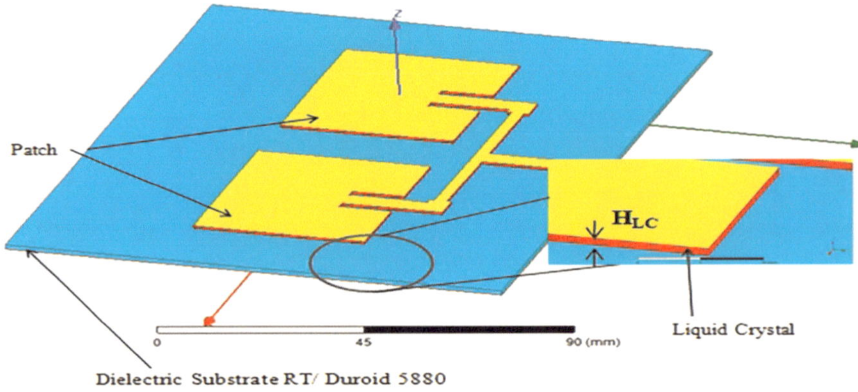

Figure 4.
Orthographic view of inset feed microstrip patch antenna array with LC substrate.

substrate with LC materials is done. LC material has excellent tuning property with moderate losses at microwave frequency. Based on this excellent property, a GHz reconfigurable patch antenna is designed for WLAN/WiMAX application.

3.1.4 Reconfigurable array antenna using graphene material

In classical antenna theory, the electrical current wave traveling along a PEC antenna propagates at the speed of light in vacuum c_0 with wave vector k_0. On the other hand, the speed of electrical current wave traveling along a graphene antenna is much slower with wave vector k_{spp}. This slow propagation of current wave is responsible for the reduction of physical antenna size in accordance with the SPP wave compression factor $\mathrm{Re}\{k_{spp}\}/k_1$. As a result of these two major differences, we can say that the main factor that differentiates the antennas at microwave frequency from the antennas at THz frequency is plasmonic effect. According to key physics, "Maxwell's equations are scale-invarient" means with scaling we can shift antenna's resonating frequency in microwave range from lower to higher frequency and vice versa with the same performance. However, when we try to apply the same principle to shift the antenna's resonating frequency from microwave frequency to THz frequency, we observe the effect of plasmonic in metals, which give the additional effects. Hence, at higher frequency, Maxwell's equations get modified and we find the "Hybrid electrical-optical solution of Maxwell's equations". Keeping these effects in mind, the antenna design procedure at THz frequency includes few steps. First, a microstrip patch antenna is designed with monolayer graphene patch on 23 μm polyimide

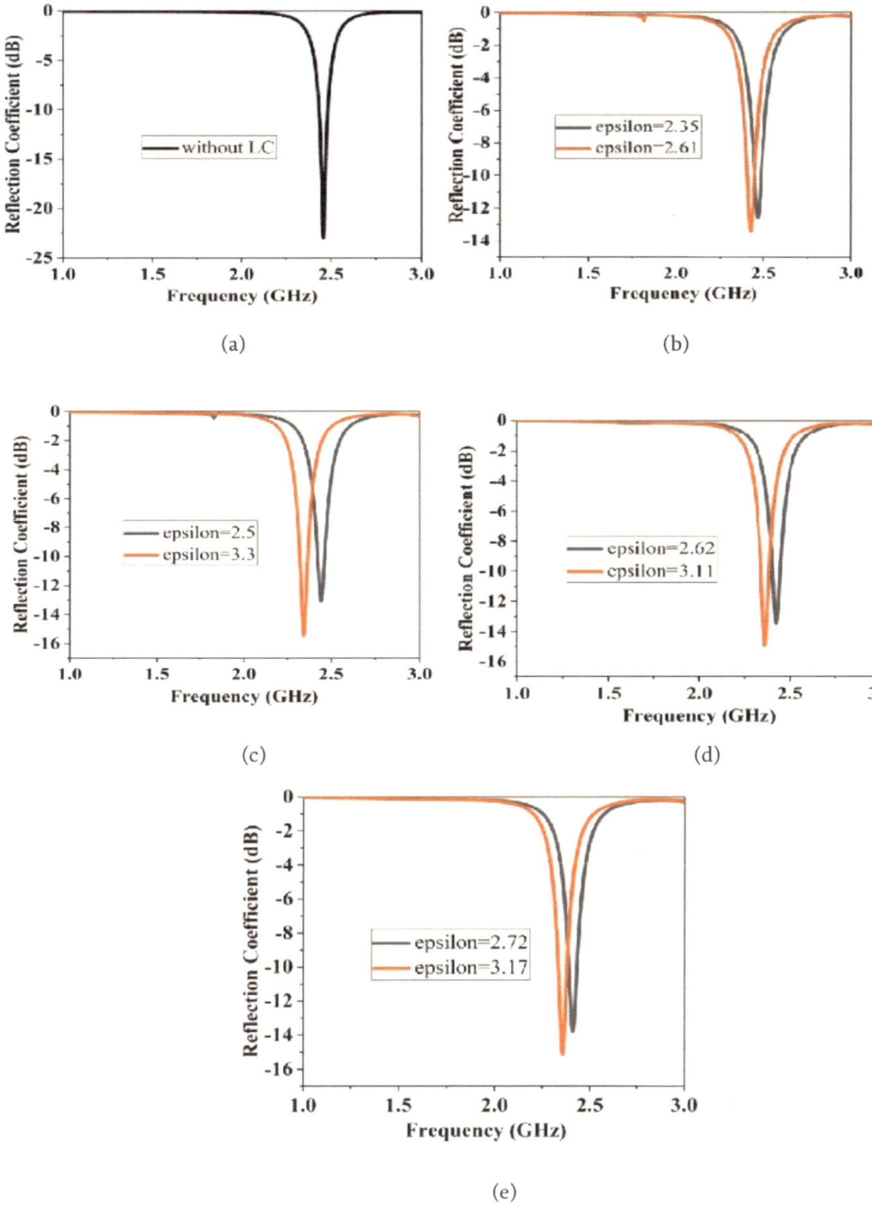

Figure 5.
Return loss of 1 × 2 antenna array (a) without LC, (b) with BL037 LC, (c) GT3-24002 LC, (d) BL006 LC, (e) E7 LC.

substrate with permittivity ε_r = 3.5 and loss tangent = 0.008. Monolayer graphene used for antenna design in this chapter exhibits fixed properties of 0.7 eV chemical potential, 1 ps relaxation time, 300 K temperature, and 0.345 graphene layer thickness. To decide the dimensions of graphene THz antenna, guidelines of conventional microstrip patch antenna are followed, but the length of the patch antenna is calculated by following Eq. (1). The feeding structure of the antenna consists of step tapered microstrip transmission line. The designed graphene antenna has attractive features such as extreme miniaturization, flexibility, and high speed. Chemical

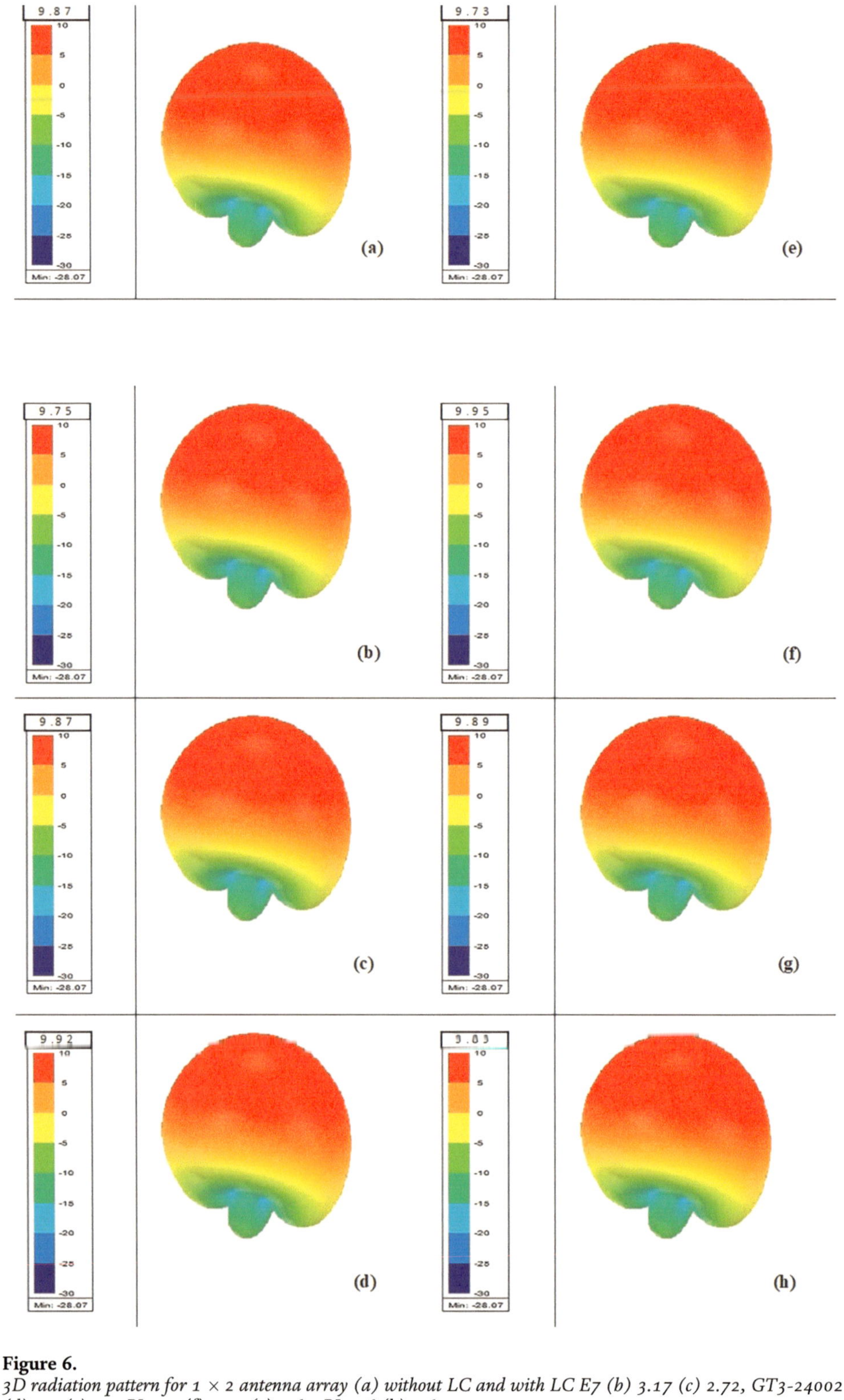

Figure 6.
3D radiation pattern for 1 × 2 antenna array (a) without LC and with LC E7 (b) 3.17 (c) 2.72, GT3-24002 (d) 2.5 (e) 3.3, BL037 (f) 2.35 (g) 2.61, BL006 (h) 2.62.

potential of graphene can be shifted above or below the Dirac point (thus altering the carrier concentration in the material) by applying a voltage. As graphene provides an extreme freedom to tune its properties, it is very convenient to make such antennas reconfigurable to achieve multifunctionality. The top and frontal views of the antenna are shown in **Figure 7(a)** and **(b)**, respectively.

$$L = m\frac{\lambda_{spp}}{2} = m\frac{\pi}{Re\{k_{spp}\}} \tag{1}$$

If the length of the graphene patch is integer multiple of half plasmon wavelengths, λ_{spp} as given in Eq. (1), the THz graphene antenna resonates, and the antenna-radiated EM field is maximized. Ultimately, the frequency response and efficiency of the antenna depend on the properties of SPP waves, which in turn depend on the conductivity of graphene [48].

Even though the graphene conductivity is high and tunable, its monoatomic thin layer sheet gives very high surface impedance due to which the gain and efficiency of graphene antennas are very low. To increase the gain and efficiency of the single element antenna, 1 × 4 antenna array is designed. Efficiency of the graphene antenna can also be improved by using multilayer structure, but for the proposed design we will work on monolayer and multielement graphene patch antenna. The orthographic view of 1 × 4 antenna array is shown in **Figure 8**.

With this 1 × 4 antenna array, not only the gain is increased but also the reconfiguration of frequency, radiation pattern, and polarization is achieved, which is demonstrated in **Figure 8**.

3.1.4.1 Design of polarization reconfigurable graphene antenna

From the properties of graphene given above, the chemical potential of graphene can be shifted above or below the Dirac point (thus altering the carrier concentration in the material) by applying a voltage. This unique tunable property of graphene can be used to design a frequency reconfigurable antenna. When a bias voltage is applied to the designed antenna array as shown in **Figure 9**, the chemical potential of graphene is varied from 0.4 eV to 0.8 eV. With the increase of chemical potential and change of surface conductivity, the dramatical change in antenna properties is observed. Increased chemical potential gives the increase in return loss and shifts of antenna resonance frequency to the higher frequencies. Polarization of the antenna defines the time-varying orientation and relative magnitude of the electric field vector when the antenna is radiating or receiving the EM waves. The polarization can be linear, circular, or elliptical. In terrestrial broadcasting and mobile communications,

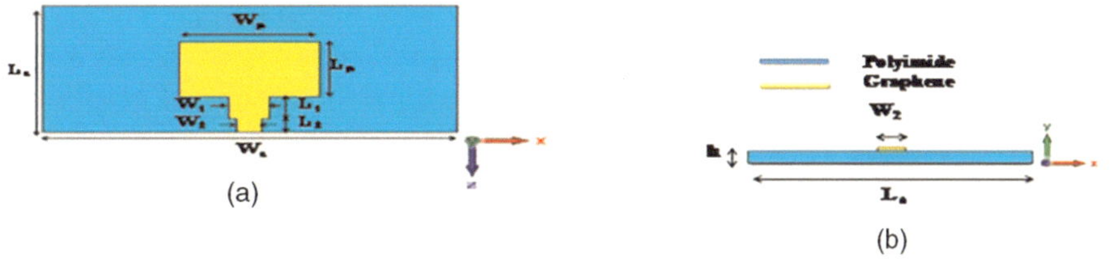

Figure 7.
(a) Top view and (b) front view of graphene-based microstrip patch antenna.

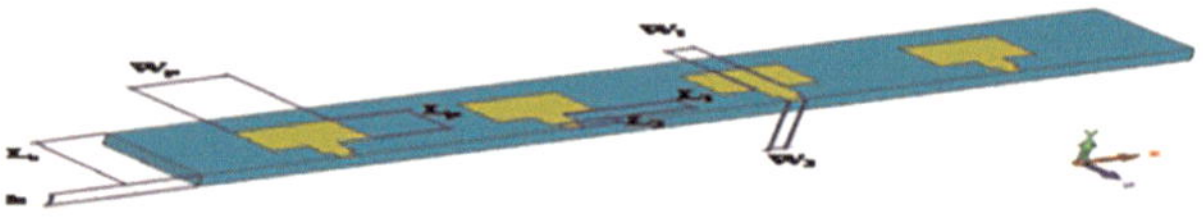

Figure 8.
Orthographic view of 1 × 4 graphene antenna array.

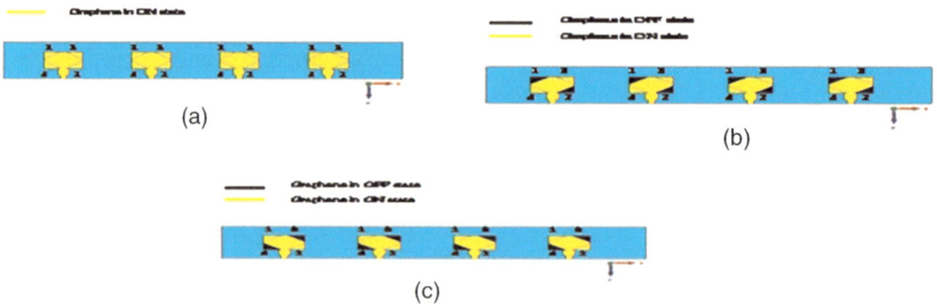

Figure 9.
(a) Model of designed LP: (b) RHCP, and (c) LHCP antenna array.

linear polarization is widely but not exclusively used as the amount of EM radiation is higher and the antennas are assumed to be aligned, while circular polarization is mainly used for satellite-to-earth communication due to the Faraday rotation effect present in the ionosphere. However, when the orientation of the antennas is unknown, circular polarization can be used as it is more robust than linear polarization in such conditions. In this section, we will analyze the use of variable conductivity of graphene as a potential method to switch between different polarizations. In the proposed structure, three scenarios are evaluated based on 1 × 4 antenna array with a rectangular graphene patch of chemical potential 0.7 eV having all four corners truncated. Graphene extensions are added to substitute the truncated corners as shown in **Figure 9**. Activating or deactivating these extensions allows one to switch between different polarizations. Linear polarizations (LPs) are achieved when all the extensions are ON, means a fixed bias voltage is applied to these corners. Right-hand circular polarization (RHCP) is set when only extensions 3 and 4 are ON, whereas left-hand circular polarization (LHCP) is achieved when only extensions 1 and 2 are ON of all the patch elements.

3.1.4.2 Design of frequency reconfigurable graphene antenna

From the properties of graphene given above, the chemical potential of graphene can be shifted above or below the Dirac point (thus altering the carrier concentration in the material) by applying a voltage. This unique tunable property of graphene can be used to design a frequency reconfigurable antenna. When a bias voltage is applied to the designed antenna array as shown in **Figure 9**, the chemical potential of graphene is varied from 0.4 eV to 0.8 eV. With the increase of chemical potential and change of surface conductivity, the dramatical change in antenna properties is observed. Increased chemical potential gives the increase in return loss and shifts of antenna resonance frequency to the higher frequencies.

3.1.4.3 Design of pattern reconfigurable graphene antenna

Novel properties of graphene easily satisfy our requirements in designing reconfigurable directional antennas for THz communication. Most of the radiation pattern reconfigurable antennas currently exist, work on the leaky-wave theory where in order to tune the radiation patterns, many different bias voltages are required. Apart from these complications, the gain of the antenna is also less. In this project, a new pattern reconfigurable antenna is designed based on graphene patch antenna array. In the given structure, we have achieved a beam reconfiguration with simple 1×4 antenna array. Directivity of this structure is pretty good. Due to monolayer graphene patch, the efficiency of the proposed structure is 60%, which can be further increased by using multilayer graphene patch. Four patch elements are used in the proposed antenna array, and an electrostatic bias voltage is applied to these patches to change its resistance modes. When the bias voltage is applied (ON state), the graphene patch is in low resistance mode, and when the bias voltage is not applied (OFF state), the graphene patch is in high resistance mode. This way with two modes and four patch elements, we can have 16 combinations. Naming the patches as A, B, C, and D as shown in **Figure 10** and denoting their OFF and ON state with 0 and 1 respectively for patch A, patch B, patch C, and patch D, we can have different combinations as shown in **Figure 11**.

Starting from the theoretical analysis, applying the structure and calculating the dimensions of the given antenna array, the simulation process is carried out in CST software. Gain of the single element monolayer graphene\-based patch antenna is 4.07 dB with efficiency 43%. To increase the gain and efficiency, we have used 1×4 antenna array. **Figure 12** shows the graph of the reflection coefficient, and **Figure 13** shows the simulated radiation pattern of the single element antenna and 1×4 antenna array in theta and phi plane. Reflection coefficient graph is plotted as a function of frequency in range of 0.2–1 THz.

With single element patch, the value of reflection coefficient obtained is -18.56 dB at resonating frequency 0.65 THz, whereas with multielement patch, the value of reflection coefficient obtained is -41.11 dB at resonating frequency 0.71 THz.

Table 4 shows the comparison of radiation characteristics for single element patch and 1×4 patch antenna.

The results presented in this section are divided into the results obtained from linear, RHCP, and LHCP reconfigurable antenna with graphene extensions in the corner. Axial ratio (AR) is a factor used to determine the polarization and represents the ratio between the major and the minor magnitudes of the electric field along each of the x- and y-axis [49]. It follows Eq. (2)

$$AR = E_x/E_y; \text{or } E_y/E_x \tag{2}$$

where E_x and E_y are the electric field components in the x- and y-axis, respectively. Ideally, the AR needed for an antenna to radiate with a circular polarization is AR = 0 dB. However, an AR within 3 dB of this value is still accepted as CP. In contrast, linear

Figure 10.
Naming of patch elements in 1×4 graphene-based antenna array.

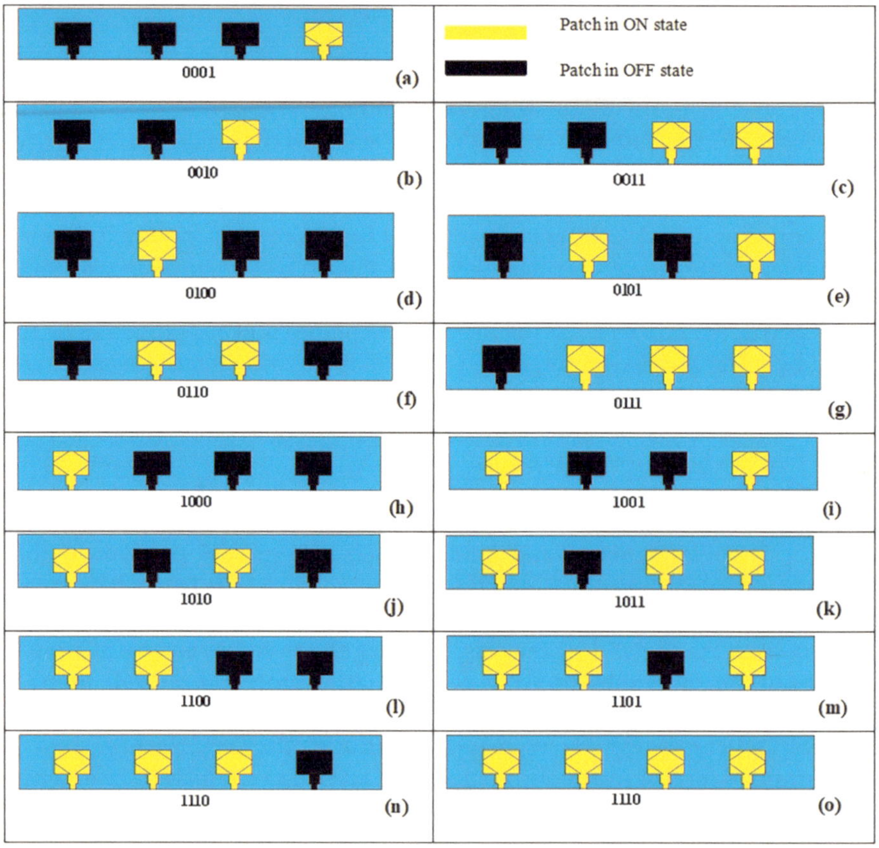

Figure 11.
Different arrangements of states of patch elements in 1 × 4 antenna array for radiation pattern reconfiguration. Depending upon switching, it is made, when switch is black, it is off condition.

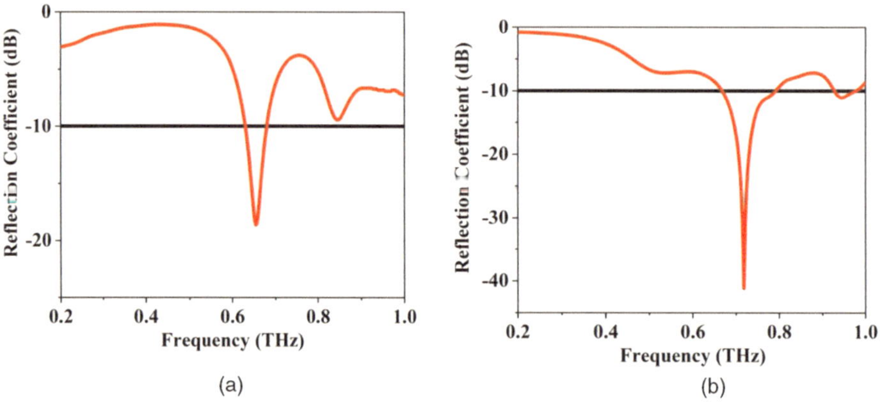

Figure 12.
Reflection coefficient of (a) single patch antenna and (b) 1 × 4 antenna array.

polarization is achieved when the AR is as high as possible (electric field is being propagated only on one of the axis). **Figure 5** demonstrate the graph of reflection coefficient and AR for differently polarized antenna. Reflection coefficient graph for

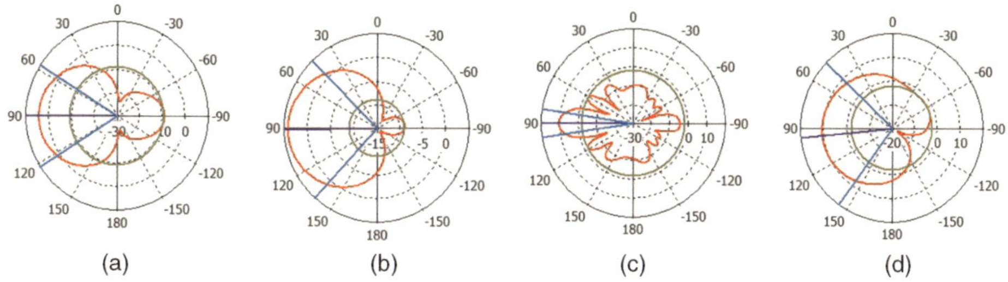

Figure 13.
2D polar plot of single patch antenna in (a) phi plane, (b) theta plane and 1 × 4 antenna array in (c) phi plane, (d) theta plane.

Parameters	Single patch	1 × 4 patch
Reflection coefficient (dB)	−18.56	−41.11
Gain (dB)	4.02	10.49
Directivity (dB)	7.57	12.69
Efficiency (%)	43	60

Table 4.
Radiation characteristics of single patch and 1 × 4 array patch antenna.

linearly and circularly polarized antenna is plotted as a function of frequency in the range of 0.2–1 THz with almost same response. Slight shift in resonating frequency is observed when polarization is changed from linear to circular. Axial ratio graph is plotted as a function of theta angle to show the polarization effect in the main beam direction. Reference line shown in the axial ratio graph represents the main beam direction of the antenna at resonating frequency. Axial ratio obtained for linearly polarized antenna and circularly polarized antenna is 30 dB and 1.58 dB, respectively.

To visualize the sense of CP radiation of the proposed antenna, simulated surface current on the patch elements for four phase angles 0° (time t = 0), 90° (t = T/4), and 270° (t = T/2) is shown in **Figure 5**. Surface current distribution of the LHCP geometry is shown in **Figure 14(a)** with the orientation of surface current from 0° to 270° at 0.515 THz. For 0° phase reference, the currents are directed to the −z direction. For 90° phase difference, the current flow is dominated by −x direction. For 180° phase angle, the current flow is oppositely directed to 0°, which is in +z direction. Again for 270° the current flow is +x directed. Similarly, the current distribution can be explained for RHCP from **Figure 14(b)**. Surface current distribution direction for the phase 0° is equal in magnitude and opposite in direction to 180°. Same is the case of surface current distribution direction at 90° and 270°. Hence, the criteria for CP are satisfied. As the direction of view is chosen as +y-axis, the direction of rotation of current is anticlockwise and the sense of polarization is confirmed as LHCP. The direction of rotation of current is clockwise and the sense of polarization is confirmed as RHCP (**Figures 15** and **16**).

Table 5 shows the summary of radiation characteristics of proposed polarization reconfigurable antenna.

To analyze the property of frequency reconfigurable antenna, the proposed structure is simulated in CST software, and the result is shown in **Figure 17**. Chemical potential of all the four patch elements of 1 × 4 antenna array are changed simultaneously from 0.4 eV to 0.8 eV. Obtained simulated result is summarized in

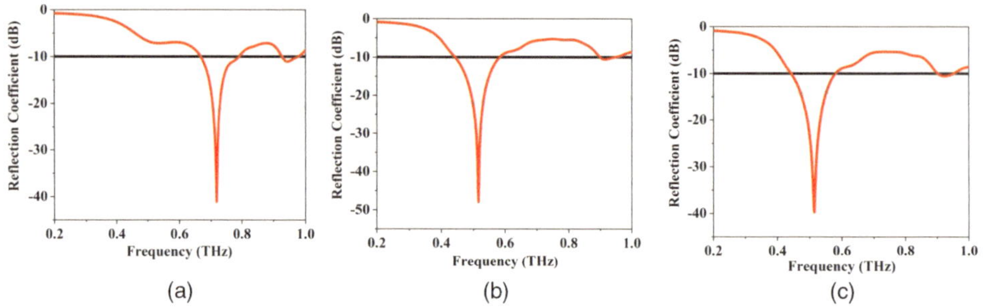

Figure 14.
Reflection coefficient of 1 × 4 antenna array for (a) LP, (b) LHCP, and (c) RHCP.

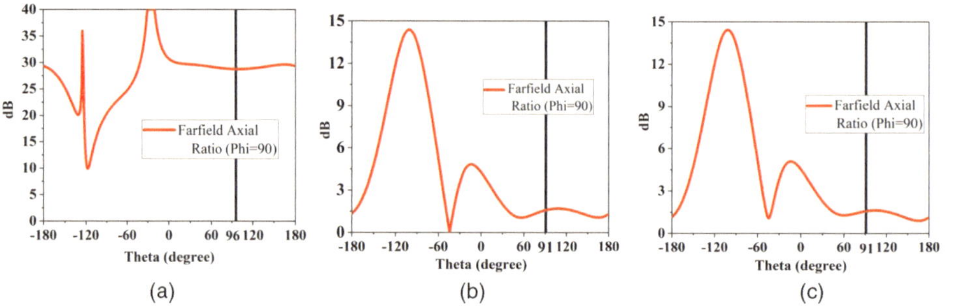

Figure 15.
Axial Ratio of 1 × 4 antenna array for (a) LP, (b) LHCP, and (c) RHCP.

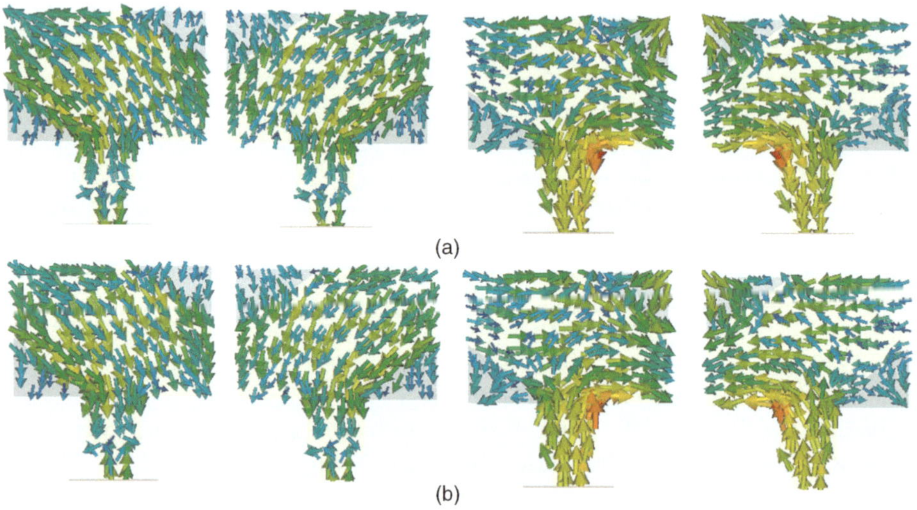

Figure 16.
Simulated surface current distributions on single patch element of the antenna array at 0.515 THz for (a) LHCP, (b) RHCP at different phases (i) 0°, (ii) 90°, (iii) 180°, and (iv) 270°.

Table 6. Shift in frequency is observed from 0.64 THz to 0.74 THz with change of chemical potential from 0.4 eV to 0.8 eV. It is clear that the achieved frequency band, which is covered by the proposed frequency reconfigurable antenna, is 0.64–0.74 THz,

Polarization	Axial ratio (dB)	Main beam direction (degree)	Gain (dBi)	Efficiency (%)
LP	30.9	96	10.5	60
LHCP	1.58	91	6.86	38
RHCP	1.58	91	6.77	38

Table 5.
Radiation characteristics of polarization reconfigurable antenna.

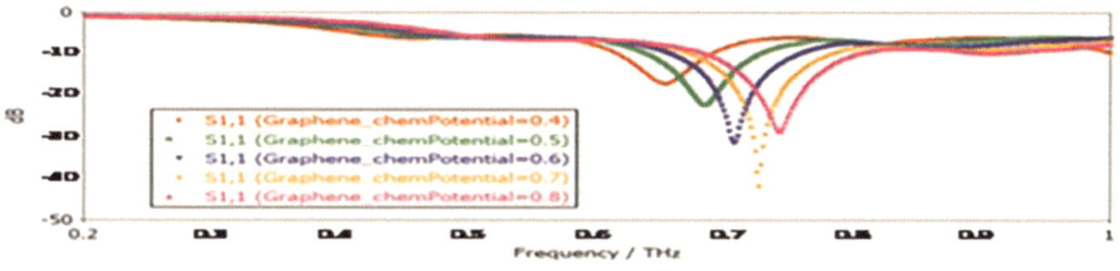

Figure 17.
Effect of variation of graphene chemical potential on resonating frequency.

which gives the bandwidth of 100 GHz, which is high enough. Hence, with such reconfigurable antennas, we can easily remove the drawback of narrow band antennas.

Table 7 depicts the simulated radiation characteristics of pattern reconfigurable antenna. The main beam direction of the antenna array can be reconfigured based on the ON/OFF state of each patch element of 1×4 antenna array as shown in **Figure 10**. With two modes (ON/OFF) and four patch elements, we can have 16 combinations, but here we can ignore the case when all patch elements are in OFF state because in this case the surface resistance of graphene will be very high for all the patches, and hence, the antenna will not radiate. **Figure 18** shows the 2D polar plot in orthogonal plane for rest of the 15 cases.

4. Conclusion

As reconfigurable antenna is very useful nowadays in the similar manner including array technique makes it more applicable in many applications such as satellite communication, biomedical applications, radar systems, smart new generation systems such as 5G and beyond, etc. Graphene, comparatively very new material as per design of antenna in terahertz regime, is most required wonder material for reconfigurable array antenna. Apart from graphene, liquid crystal is also good material to design this antenna in gigahertz range. Also, there are so many other techniques to design and make it useful for various applications. Reconfigurable antennas play important roles in smart and adaptive systems and are the subject of many research studies. They offer several advantages such as multifunctional capabilities, minimized volume requirements, low front-end processing efforts, and good isolation; these makes them well suited to use in wireless applications such as fifth-generation (5G) mobile terminals. There are so many techniques available to reconfigure the antenna properties such as electrical, optical, mechanical, and material-based reconfiguration. Among all the options, material-based reconfigurable antennas are the simplest ones with excellent performance. Where liquid crystal is proved to be suitable candidate in microwave frequency range, the 2D material like graphene is proved to be a real gem in THz frequency range.

Patch A	Patch B	Patch C	Patch D	Main beam direction in phi plane (degree)	Gain (dBi)
OFF	OFF	OFF	ON	104	5.61
OFF	OFF	ON	OFF	113	5.03
OFF	OFF	ON	ON	82	6.8
OFF	ON	OFF	OFF	115	4.92
OFF	ON	OFF	ON	89	6.19
OFF	ON	ON	OFF	90	6.85
OFF	ON	ON	ON	89	8.95
ON	OFF	OFF	OFF	75	5.45
ON	OFF	OFF	ON	65	6.9
ON	OFF	ON	OFF	91	6.21
ON	OFF	ON	ON	90	8.86
ON	ON	OFF	OFF	100	6.82
ON	ON	OFF	ON	90	8.78
ON	ON	ON	OFF	91	8.8
ON	ON	ON	ON	90	10.4

Table 6.
Performance parameters of radiation pattern reconfigurable antenna.

Graphene chemical potential (eV)	Frequency (THz)
0.4	0.64
0.5	0.69
0.6	0.70
0.7	0.71
0.8	0.74

Table 7.
Performance parameters of frequency reconfigurable antenna.

Frequency reconfiguration is explained and polarization reconfiguration is discussed with linear, right-hand circular, and left-hand circular polarization. Radiation pattern reconfiguration is also explained with beam variation range of 82°–115°. Efficiency of the proposed structure is slightly less because monolayer graphene has high surface resistance due to very high surface area to volume. But this issue can be eliminated by using multilayer graphene. Hence, graphene-based antennas offer better and simpler reconfiguration technique than any other existing technique at THz frequency.

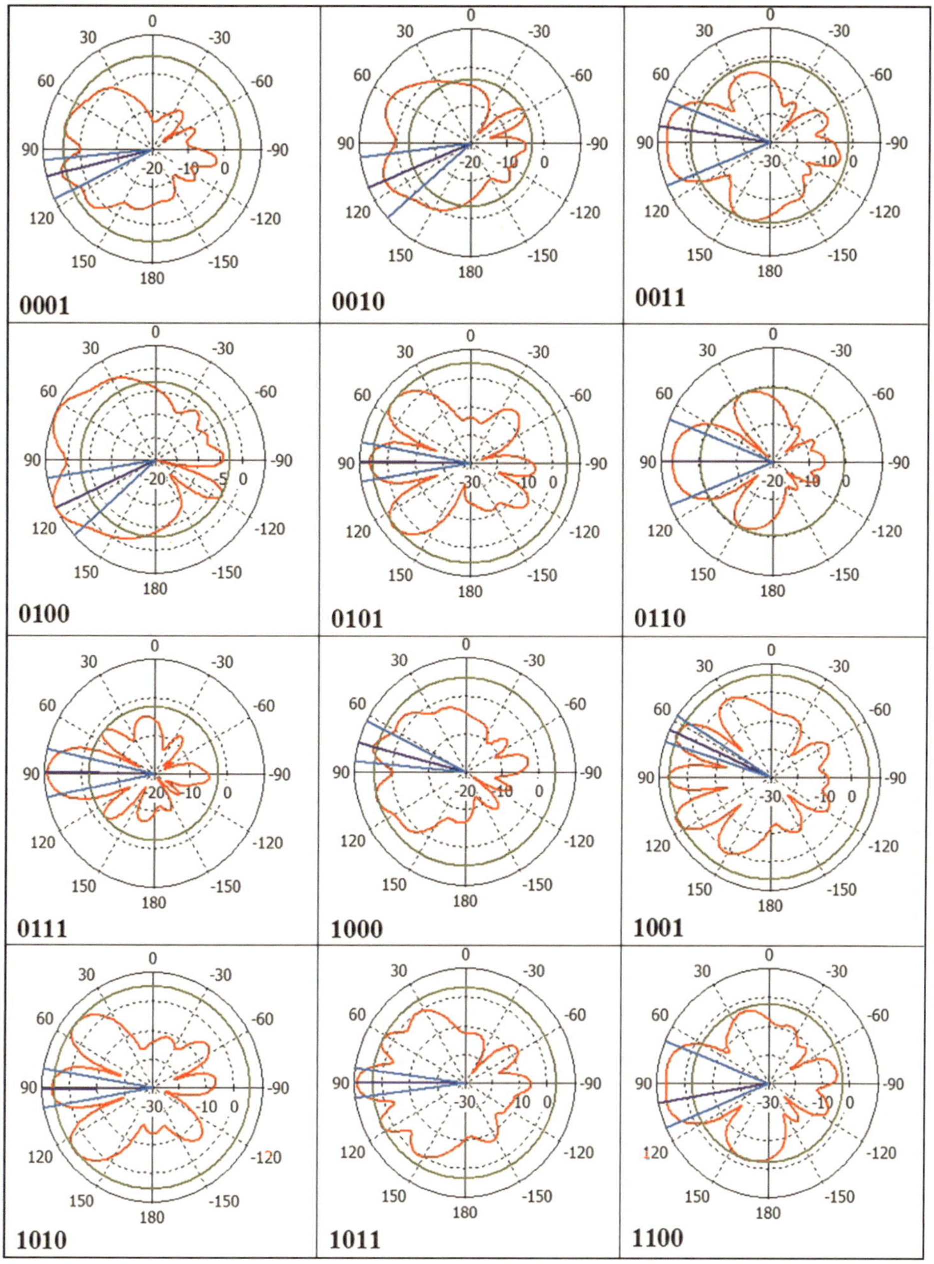

Figure 18.
2D polar plot of pattern reconfigurable antenna in orthogonal plane with different arrangements of patch elements.

Author details

Smrity Dwivedi
Department of Electronics Engineering, IIT BHU, Varanasi, India

*Address all correspondence to: sdwivedi.ece@itbhu.ac.in

References

[1] Balanis CA. Antenna Theory: Analysis and Design. John Wiley & Sons; 2016

[2] Balanis CA, editor. Modern Antenna Handbook. John Wiley & Sons; 2011

[3] Schaubert D. Frequency-agile polarization diversity microstrip antennas and frequency scanned arrays. U.S. Patent 4. 1983. pp. 367-474

[4] Christodoulou CG, Tawk Y, Lane SA, Erwin SR. Reconfigurable antennas for wireless and space applications. Proceedings of the IEEE. 2012;**100**: 2250-2261

[5] Mohanta HC, Kouzani AZ, Mandal SK. Reconfigurable antennas and their applications. Universal Journal of Electrical and Electronic Engineering. 2019;**6**:239-258

[6] Bucci OM, Mazzarella G, Panariello G. Reconfigurable arrays by phase-only control. IEEE Transactions on Antennas and Propagation. 1991;**39**:919-925

[7] Dürr M, Trastoy A, Ares-Pena F. Multiple pattern linear antenna arrays with single prefixed amplitude distributions: Modified Woodward-Lawson synthesis. Electronics Letters. 2000;**36**:1345-1346

[8] Díaz X, Rodríguez-González JA, Ares-Pena F, Moreno E. Design of phase-differentiated multiple-pattern antenna arrays. Microwave and Optical Technology Letters. 2000;**16**:52-53

[9] Ares-Pena FJ, Franceschetti G, Rodríguez-González JA. A simple alternative for beam reconfiguration of array antennas. Progress In Electromagnetics Research. 2008;**88**: 227-240

[10] Rodríguez-González JA, Trastoy A, Brégains JC, Ares-Pena FJ, Franceschetti G. Beam reconfiguration of linear arrays using parasitic elements. Electronics Letters. 2006;**42**:3

[11] Yagi H. Beam transmission of ultra-short waves. Proceedings of the IRE. 1928;**16**:715

[12] Yuan HW, Gong SX, Zhang PF, Wang X. Wide scanning phased array antenna using printed dipole antennas with parasitic element. Progress in Electromagnetics Research. 2008;**2**: 187-193

[13] Elliott RS. An Introduction to Guided Waves and Microwave Circuits. Englewood Cliffs, NJ, USA: Prentice-Hall; 1993

[14] Smith EK, Weintraub S. The constants in the equation for atmospheric refractive index at radio frequencies. Proceedings of the IRE. 1953;**41**:1035-1037

[15] Salas-Sánchez AA, López-Martín ME, Rodríguez-González JA, Ares-Pena FJ. Design of polyimide-coated Yagi-Uda antennas for monitoring the relative humidity level. IEEE Transactions on Geoscience and Remote Sensing. 2017;**14**:961-963

[16] Sihvola AH. How strict are theoretical bounds for dielectric properties of mixtures? IEEE Transactions on Geoscience and Remote Sensing. 2002;**40**:880-886

[17] Brown ER. RF-MEMS switches for reconfigurable integrated circuits. IEEE Transactions on Microwave Theory and Techniques. 1998;**46**(11):1868-1880

[18] Erdil E, Topalli K, Unlu M, Civi OA, Akin T. Frequency tunable microstrip patch antenna using RF MEMS technology. IEEE Transactions on Antennas and Propagation. 2007;**55**(4): 1193-1196

[19] Peroulis D, Sarabandi K, Katehi LPB. Design of reconfigurable slot antennas. IEEE Transactions on Antennas and Propagation. 2005;**53**(2):645-654

[20] Pagri TG. Design and implementation of reconfigurable antenna. International Research Journal of Engineering and Technology (IRJET). 2017;**4**:2655-2660

[21] Oh SS, Jung YB, Ju YR, Park HD. Frequency-tunable open-ring microstrip antenna using varactor. In: 2010 International Conference on Electromagnetics in Advanced Applications. IEEE; 2010. pp. 624-626

[22] Panagamuwa CJ, Chauraya A, Vardaxoglou JC. Frequency and beam reconfigurable antenna using photoconducting switches. IEEE Transactions on Antennas and Propagation. 2006;**54**(2):449-454

[23] Costantine J, Tawk Y, Barbin SE, Christodoulou CG. Reconfigurable antennas: Design and applications. Proceedings of the IEEE. 2015;**103**: 424-437. DOI: 10.1109/JPROC.2015. 2396000

[24] Missaoui S, Gharbi A, Kaddour M. Design and simulation reconfigurable liquid crystal patch antennas on foam substrate. Journal of Chemical Engineering and Materials Science. 2011; **2**(31):96-102

[25] Papanicolaou NC, Christou MA, Polycarpou AC. Frequency-agile microstrip patch antenna on a biased liquid crystal substrate. Electronics Letters. 2015;**51**(5):202-204. DOI: 10.1049/el.2014.3856

[26] Polycarpou AC, Christou MA, Papanicolaou NC. Tunable patch antenna printed on a biased nematic liquid crystal cell. Institute of Electrical and Electronics Engineers. 2014;**62**(30): 4980-4987. DOI: 10.1109/TAP.2014. 2344099

[27] Liu L, Langley RJ. Liquid crystal tunable microstrip patch antenna. Electronics Letters. 2008;**44**(25): 1179-1180. DOI: 10.1049/el:20081995

[28] Missaoui S, Missaoui S, Kaddour M. Tunable microstrip patch antenna based on liquid crystals. In: Direct and Inverse Problems of Electromagnetic and Acoustic Wave Theory (DIPED), 2016 XXIst International Seminar/Workshop on. Vol. 51. IEEE; 2016. pp. 88-91. DOI: 10.1109/DIPED.2016.7772222

[29] Martin N, Laurent P, Person C, Gelin P, Huret F. Patch antenna adjustable in frequency using liquid crystal. In: 33rd European Microwave Conference Proceedings (IEEE Cat. No. 03EX723C). Vol. 2. IEEE; 2003. pp. 699-702

[30] Zhao Y, Huang C, Qing AY, Luo X. A frequency and pattern reconfigurable antenna array based on liquid crystal technology. IEEE Photonics Journal. 2017;**9**(3):1-7

[31] Pan C, He Z, Liu Y. A novel frequency reconfigurable microstrip antenna based on liquid crystal. In: Proceedings of the 2nd International Conference on Telecommunications and Communication Engineering. ACM; 2018. pp. 201-204

[32] Sethi GS, Srivastava A, Chiu CY, Chigrinov V, Murch RD. Design of a transparent LC based reconfigurable

antenna. In: 2016 Asia-Pacific Microwave Conference (APMC). IEEE; 2016. pp. 1-4

[33] Yasir M, Savi P, Bistarelli S, Cataldo A, Bozzi M, Perregrini L, et al. A planar antenna with voltage-controlled frequency tuning based on few-layer graphene. IEEE Antennas and Wireless Propagation Letters. 2017;**16**:2380-2383

[34] Leng T, Huang X, Zhang X, Hu Z. Reconfigurable dipole antenna design using graphene based switch. In: 2015 IEEE International Symposium on Antennas and Propagation & USNC/ URSI National Radio Science Meeting. IEEE; 2015. pp. 2295-2296

[35] Huang Y, Wu LS, Tang M, Mao J. Design of a beam reconfigurable THz antenna with graphene-based switchable high-impedance surface. IEEE Transactions on Nanotechnology. 2012; **11**(4):836-842

[36] Goyal R, Vishwakarma DK. Design of a graphene-based patch antenna on glass substrate for high-speed terahertz communications. Microwave and Optical Technology Letters. 2018;**60**(7): 1594-1600

[37] Dash S, Patnaik A. Graphene loaded frequency reconfigurable metal antenna. In: 2017 IEEE International Conference on Antenna Innovations & Modern Technologies for Ground, Aircraft and Satellite Applications (iAIM). IEEE; 2017. pp. 1-4

[38] Wu Y, Qu M, Jiao L, Liu Y, Ghassemlooy Z. Graphene-based Yagi-Uda antenna with reconfigurable radiation patterns. AIP Advances. 2016;**6** (6):065308

[39] Dash S, Patnaik A, Kaushik BK. Performance enhancement of graphene plasmonic nanoantennas for THz communication. IET Microwaves, Antennas & Propagation. 2018;**13**(1): 71-75

[40] Naghdehforushha SA, Moradi G. High directivity plasmonic graphene-based patch array antennas with tunable THz band communications. Optik. 2018; **168**:440-445

[41] Tamagnone M, Gomez-Diaz JS, Mosig JR, Perruisseau-Carrier J. Reconfigurable terahertz plasmonic antenna concept using a graphene stack. Applied Physics Letters. 2012;**101**(21): 214102

[42] Zhang X, Ruan C, Dai J, ul Haq T. Frequency and radiation pattern reconfigurable graphene square spiral antenna at terahertz band. In: 2018 IEEE Asia-Pacific Conference on Antennas and Propagation (APCAP). IEEE; 2018. pp. 1-2

[43] Perez-Palomino G, Barba M, Encinar JA, Cahill R, Dickie R, Baine P, et al. Design and demonstration of an electronically scanned reflectarray antenna at 100 GHz using multiresonant cells based on liquid crystals. IEEE Transactions on Antennas and Propagation. 2015;**63**(8): 3722-3727

[44] Chandra S, Dwivedi S. Comparative analysis of reconfigurable patch antenna array for different liquid crystal substrates. In: URSI AP-RASC 2019, 09-15 March 2019, New Delhi, India. IEEE Explore

[45] Mrunalini S, Manoharan A. Dual-band re-configurable graphene-based patch antenna in terahertz band for wireless network-on-chip applications. IET Microwaves, Antennas & Propagation. 2017;**11**(14): 2104-2108

[46] Chandra S, Dwivedi S. Graphene based radiation pattern reconfigurable antenna. In: International Conference for Convergence in Technology. IEEE explore; April 2-4, 2021. DOI: 10.23919/URSIAP-RASC.2019.8738497

[47] Karabey OH. Electronic Beam Steering and Polarization Agile Planar Antennas in Liquid Crystal Technology. Springer Science & Business Media; 2013

[48] Jornet JM, Akyildiz IF. Graphene-based plasmonic nano-antenna for terahertz band communication in nanonetworks. IEEE Journal on Selected Areas in Communications. 2013;**31**(12): 685-694

[49] Alvarez CN, Cheung R, Thompson JS. Polarization reconfigurable antennas using graphene for microwave applications. In: 2015 IEEE International Conference on Ubiquitous Wireless Broadband (ICUWB). IEEE; 2015. pp. 1-5